Optimal Bundling

Marketing Strategies
for Improving
Economic Performance

Springer
Berlin
Heidelberg
New York
Barcelona
Hong Kong
London
Milan
Paris
Singapore
Tokyo

Ralph Fuerderer · Andreas Herrmann
Georg Wuebker (Eds.)

Optimal Bundling

Marketing Strategies for Improving Economic Performance

With 37 Figures
and 48 Tables

Dr. Ralph Fuerderer
Adam Opel AG
International Technical Development Center
65428 Ruesselsheim
Germany

Prof. Dr. Andreas Herrmann
University of Mainz
School of Business Administration
55099 Mainz
Germany

Dr. Georg Wuebker
Simon, Kucher & Partners
Haydnstraße 36
53115 Bonn
Germany

ISBN 3-540-65247-7 Springer-Verlag Berlin Heidelberg New York

Library of Congress Cataloging-in-Publication Data
Die Deutsche Bibliothek – CIP-Einheitsaufnahme
Optimal bundling: marketing strategies for improving economic performance: with 48 tables / Ralph Fuerderer, Andreas Herrmann, Georg Wuebker (eds.). – Berlin; Heidelberg; New York; Barcelona; Hong Kong; London; Milan; Paris; Singapore; Tokyo: Springer, 1999
ISBN 3-540-65247-7

Printed in Germany

Hardcover-Design: Erich Kirchner, Heidelberg

SPIN 10700610 42/2202-5 4 3 2 1 0 – Printed on acid-free paper

Acknowledgements

We would like to thank the authors, Richard Adam, Hans H. Bauer, Andreas Cornet, Robin Higie Coulter, Charley Fine, Frank Huber, Arnd Huchzermeier, Volker Lingnau, Vijay Mahajan, Kent B. Monroe, Carl-Stefan Neumann, Jürgen Ringbeck, Linus Schrage, Hermann Simon, Rajneesh Suri, Nils Tönshoff, Martin Wricke, and Manjit S. Yadav for their contributions. We offer special thanks to Dipl.-Kff. Christine Braunstein and Dominic Hinzer, who helped us coordinate the production of the book. At Springer, Werner A. Mueller provided financial support and encouraged us to get the book ready.

Ralph Fuerderer
Andreas Herrmann
Georg Wuebker

Contents

Part 1: Introduction

Part 2: Optimization Approaches

Part 3: Behavioral Aspects

Part 1: Introduction

Introduction to Price Bundling

Ralph Fuerderer[1], Andreas Herrmann[2], and Georg Wuebker[3]

[1] **Raph Fuerderer**, Adam Opel AG, International Technical Development Center, Ruesselsheim, Germany.
[2] **Andreas Herrmann**, Johannes Gutenberg-University of Mainz, Germany.
[3] **Georg Wuebker**, SIMON, KUCHER & PARTNERS, Bonn, Germany.

Collecting goods or services in a package and selling them at a (discounted) package price has become a widespread sales practice in many production or service oriented industries. The methodology itself is called Price Bundling and can be classified as a price differentiation tool for the price management of multi-product/service enterprises.

The following examples demonstrate on one hand the versatility of price bundling. On the other hand, they may verify the observation that the practice of price bundling goes unnoticed many times despite its dissemination:

A recent example in the German micro-computer branch is Vobis Microcomputer AG. Their rigorous bundling strategy was considered to be an important factor of their market leadership. Each customer can build his individual Vobis bundle, which consists of tower, monitor, printer, and additional hard- and software. Bundle discounts at Vobis generally range from 5 to 10 %, and relatively increase with the bundle price. McDonald`s Value Meals combine main courses, side orders, and drinks. Compared to the individually priced ingredients, the customer saves up to 15% by purchasing the menu. High customer acceptance not only increases overall consumption. Overall profitability is also affected through the inclusion of low-cost items (french-fries, sodas), and thus may by far outweigh the bundle discount.

In the automobile industry, a car model is offered along with features such as air-conditioning, sun-roof, navigation system, and so forth. If these features are not part of the standard car offer (standard options), they can either be purchased separately (free-flow options) or within bundles (equipment packages). The car manufacturer can furthermore control the sales volume of a particular feature (hence manufacturing and purchasing costs) by applying a smart bundling strategy.

Recently, even cross-industry bundling can be observed. As one of many examples, German Lufthansa offer the Lufthansa Card in alliance with credit institutes Eurocard and VISA, German telecomunication giant Telekom, and many hotel and car rental companies. This multi-purpose card not only includes air transportation and credit card services, but furthermore a telephone chip, accident insurance and special hotel and car rental conditions. Swiss army-knives come in different models (Camper, Sportsman, Climber, Waiter etc.) containing a set of common tools (blade, cork-screw, can-opener, etc.) plus a set of customized tools (scissors, wood-saw, carrying-hook, etc.). Not only is a product bundle the only feasible design to sell this product, bundling even more adds to its unique character.

The decision of a multi-product enterprise, whether to price its goods/services individually, whether to offer them within bundles, or whether to choose a mixed form of the above is mostly not a straightforward one. Some crucial questions regarding a price bundling approach could be the following:

- What is the potential benefit of price bundling versus single pricing?
- How does price bundling fit into a company`s sales strategy?
- How can one decide upon the design and price of a bundle?
- What data is needed and how can it be collected?
- Which aspects of customer`s purchase behavior must be accounted for?

Some of these questions have been discussed in economic literature for about 30 years. However, many results have only been of low relevance for the decision process in managerial practice and have treated the problem under quite specific views and conditions.

With the book in hand, we seek to draw a comprehensive picture of most recent topics about price bundling. We elucidate all relevant aspects about the decision process on the seller`s side, as well as on the buyer`s side. The book enables practitioners to improve the expected profitability or their product lines significantly, it is essential reading for researchers and students in the field of Production/Operations and Marketing.

The book is organized as follows: The chapters in part 1 describe price bundling as a tool for gaining competitive advantage. After the introduction to price bundling, Simon and Wuebker explain a method for exploiting profit potential.

The chapters in part 2 focus on optimization approaches, such as the model developed by Fuederer, Huchzermeier, and Schrage as well as the approach proposed by Toenshoff, Fine, and Huchzermeier. Fuerderer provides an overview of theories and methods in price bundling literature. Ringbeck, Neumann, Comet and Lingnau extend the discussions in their papers to costs of complexity and product variety. The following chapter by Wuebker and Mahajan examines the use of conjoint analysis to measure reservation prices and to create price bundles.

The chapters in part 3 explore behavioral aspects of price bundling. Suri and Monroe analyze consumers′ purchase intention and their evaluation of savings on product bundles. An approach for evaluating bundling strategies is presented by Wuebker, Mahajan, and Yadav. Yadav suggest a model of anchoring and adjustment for explaining buyers′ evaluation of product bundles. In the following chapters Herrmann, and Wricke describe how consumers' evaluate multidimensional prices. Approaches for creative product bundles are presented by Herrmann, Huber, and Coulter as well as Bauer, Huber, and Adam.

Bundling - A Powerful Method to Better Exploit Profit Potential

Hermann Simon[1], and Georg Wuebker[2]

[1] **Hermann Simon**, SIMON, KUCHER & PARTNERS, Bonn, Germany and Cambridge, Massachusetts.
[2] **Georg Wuebker,** SIMON, KUCHER & PARTNERS, Bonn, Germany.

1 Introduction

Most firms are multi-product companies faced with the decision whether to sell products (goods and/or services) separately at individually determined prices or whether combinations of products should be marketed in the form of "bundles" for which a "bundle price" is asked. Bundling plays an increasingly important role in many industries and some companies even build their business strategies on bundling. A renowned case is Microsoft. By smartly combining its application software into the "Office" bundle, Microsoft extended the quasi-monopoly of Word to Excel, Access and PowerPoint. Microsoft increased the market share of PowerPoint and Access when it bundled these two less attractive components with the attractive components Excel and Winword. The so-called "Office" bundle and its software components represent the standard in the application software market and have a share over 80 percent (see Wuebker, 1998, 189). In a similar way, Microsoft is trying to monopolize the Web-browser market. In mid-1996, its Web-browser Internet Explorer had a market share of 7 percent, while Netscape's Navigator had a quasi-monopoly (over 80 percent market share). By bundling the Explorer with its operating system Windows, Microsoft's share rose to 38 percent and Netscape's lead in the browser market dropped to 58 percent by early 1997. Many software experts believe that Microsoft will soon get the leadership in the Web-browser market.

2 Bundling in Practice

In practice, bundling is very popular and appears in different forms:

- In the film industry, "block booking" is frequently applied (Stigler, 1963). The distributor does not offer single films to the movie operators - from which these would most likely select only the attractive titles - but supplies them with a "block" or bundle of more and less attractive films.

- In industries like computers, machinery, contracting, etc., whole systems (e.g. central processing unit, monitor, printer, software) are sold at a "system price".

- Bundling is particularly popular in the service sector. Examples are vacation packages (airline ticket, hotel accommodation plus rent-a-car), insurance packages, or menus in restaurants (hors d'oeuvre, entree, dessert). In the fast food industry, "value meals" are heavily sold at special discounts. McDonald's value meals contain three fast-food-items, e.g. Big Mac (DM 4.95), French Fries (DM 2.00), and a drink (DM 2.50). The bundle of the three items is sold at DM 7.99, a discount of 15 percent to the sum of the individual prices.

- Publishing companies offer their advertising customers so-called "title combinations". If companies advertise in several magazines of the same publisher they get substantial rebates.

- Bundling is also very often employed in the software industry, where it is known as suite selling. Microsoft's "Office" bundle contains three software programs in the standard version (Word, Excel, and PowerPoint), which cost DM 699 each if purchased individually. The suite of three costs DM 1,099, a discount of 48 percent relative to the sum of the individual prices. The key here is to use the strength in one product to attract customer attention to other items in the company's product line.

- Car manufacturers offer packages of popular options at substantial discounts from individual rates. In 1995, Ford offered a "Transit Plus" bundle for its vans comprising ABS, power windows, one-touch locking, airbags, and other features for DM 1,500. The individual prices of these components add up to DM 4,220, so that the bundle discount amounts to 64.5 percent.

- In the telecommunication market, bundling of different products and services will become increasingly important as bundles will simplify customers' lives. A good example is Frontier Corporation. The telecommunication company offers different services (e.g. local, long-distance, and cellular phone service) in a single package, and tallied on one monthly bill.

Most of the above examples concern complementary goods or services. While complementarity certainly fosters the advantage of bundling, it is not a necessary condition for the application of this method. Usually, the bundle price is lower than the sum of the separate prices. Table 1 shows an example from Vobis, the largest computer retailer in Europe. The bundle price is between 7.5 percent and 10.4 percent below the sum of separate prices.

Table 1. Price differences between the sum of separate prices and the bundle price: an example of personal computers

Bundle offers	Sum of separate prices (in DM)	Bundle price (in DM)	Price difference in %
Sky Mini 97 Pentium 150 + HP Deskjet printer	3,652	3,398	7.5
Sky Mini 97 Pentium 166 + HP Deskjet printer	3,452	3,198	7.9
Sky Mini 97 Pentium 150 + Lexmark printer	3,973	3,598	10.4

It should be noted that the discount is not a necessary condition for bundling. If the individual products alone offer little benefit, the bundle price can be higher than the sum of separate prices. This form of bundling is known in the literature as premium bundling (Cready, 1991). Examples are collection sets where the complete set is much more expensive than individual components of the set. If customers are not well informed about a product or do not want to buy individual components and put them together they may be inclined to pay more for the system.

3 Implementation Forms of Bundling

Based on the literature and the above examples, we systemize and distinguish several implementation forms of bundling:

<u>Separate pricing:</u> The products are offered and priced individually. This form of pricing is also called pure component pricing (Adams/Yellen, 1976) or no bundling (Dolan/Simon, 1996).

<u>Pure bundling:</u> Only the bundle is sold. The products cannot be bought individually. An example is block booking in the movie industry (Stigler, 1963). Instead of giving a theater owner the choice to buy the most attractive movies, often only combined assortments of movies are offered by the movie rental companies.

<u>Mixed bundling:</u> Here, both the bundle and the individual products are offered, i.e. a combination of separate pricing and pure bundling. Normally, prices are set for the individual products and the bundle (so-called “mixed-joint bundling”, Guiltinan, 1987). McDonald’s uses this form for its value meals. Alternatively, a discount is given for the second product if the full price is paid for the first (the “leader product”). This is the so-called “mixed-leader bundling” (Guiltinan, 1987). A company in Germany is applying “mixed-leader bundling” for veterinary products. The “leader product” is priced high and innovative, whereas the second product is priced low and mature. If the company did not give a discount for the second product, the company could not sell the "leader product" at a profitable price. Drugs are often sold to hospitals in a similar manner in Germany or to Health Maintenance Organizations in the US. Mixed bundling is the most frequently used form of bundling in practice.

<u>Tie-in sales:</u> A special form of bundling are the so-called tie-in sales (Burstein, 1960). The buyer of the main product (tying good) agrees to buy one or several complementary goods (tied goods) - which are necessary to use the tying good - exclusively from the same supplier. Often the tying good is a durable (e.g., a machine, a copier, a computer) while the tied goods are non-durables such as toner, paper, X-ray material etc. Tie-in sales can be interpreted as a form of nonlinear pricing (in the sense of a two-part tariff, Phlips, 1989) as well as a form

of pure bundling (Tacke, 1989). In the former case, the customer buys one unit (so-called "yes:no-case", Simon, 1989, 14) of the tying good and several units (so-called "variable quantity-case", Simon, 1989, 14) of the tied good. In the latter case, he buys just one unit (combination of two yes:no-cases) of the tying and the tied good.

A famous example of tie-in sales is the former IBM pricing scheme for punch cards (Telser, 1965; Warhit, 1980; Liebowitz, 1983; Simon, 1992; Dolan/Simon, 1996, 232). IBM maintained a quasi-monopoly on tabulating machines and required their customers to buy all punch cards (on which IBM had no monopoly) from IBM. In the 1930's punch cards accounted for more than 30 percent of IBM's overall annual revenues (Sobel, 1981; Simon et al., 1995). By tying in punch card sales to machine sales, IBM extended its monopoly from machines to cards. A different interpretation of this pricing strategy is that the consumption of cards measures the utilization of a machine and, thus, its value to the customer (Telser, 1965). The more cards a customer uses, the more valuable a machine is to him or her. Through the tie-in system, IBM implemented a price differentiation across customers according to their intensity of machine utilization. If the customers were allowed to buy the cards from other suppliers, the value differentials would have to be "skimmed" solely through the machine price. Accordingly, the machines would have had to be much more expensive and a price differentiation would have been difficult to implement. While the customers decide on their consumption of punch cards, they actually determine the "total" price (or bundle price) they pay for the tabulation process. In this way, they reveal the value they derive from tabulation. Other well-known cases of tie-in sales discussed in the literature are Xerox (Blackstone, 1975; Nagle, 1987; Phlips 1989, Simon et al., 1995), International Salt (Peterman, 1979) or Northern Pacific (Cummings and Ruther, 1979).

<u>Add-on bundling:</u> This special form of pricing is not strictly a mixed bundling option, because an add-on product will not be sold unless the lead product is purchased (Guiltinan, 1987, 84). Therefore, add-on bundling is similar to tie-in sales (Wuebker, 1998). Examples of this pricing strategy are a car wash (i.e., the lead service) bundled with extra wax (i.e, the add-on service) or a basic car model (i.e., the lead product) bundled with special features like ABS, power windows or airbags.

Sales rebates: Companies frequently offer customers a year-end rebate on total annual sales across all the company's products. These bonuses are mainly aimed at increasing customer loyalty (Dolan/Simon, 1996). Overall, sales bonuses are a mixture between bundling and nonlinear pricing (Wilson, 1993; Wuebker, 1998), because it does not matter whether the total sales come from one or from several products.

Cross couponing: U.S. consumer-goods manufacturers frequently use coupons to promote other products from their assortment. Coca-Cola, for example, reached diet soft-drink users by distributing coupons for Diet Minute Maid on its 2-liter Diet Coke bottle (Foster, 1991). Cross-couponing is often used to introduce new products and/or increase the sales of weak products by linking them with established products in the firm's product line. This is structurally similar to bundling in that a customer of both Diet Coke and Diet Minute Maid willing to redeem the coupon would pay less than two individuals each buying one of the two flavors.

Figure 1 illustrates the different implementation forms of bundling.

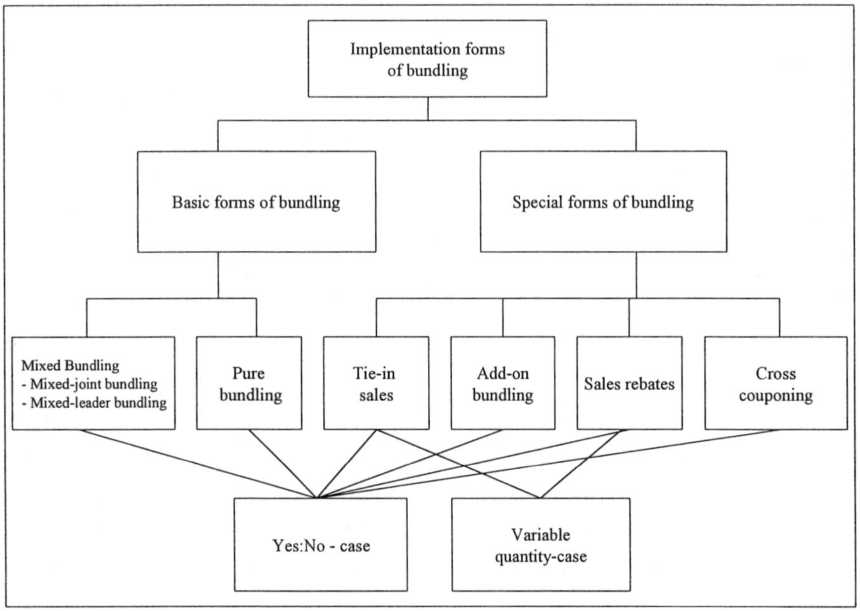

Fig. 1. Implementation forms of bundling

4 The Rationales for Bundling

Why is bundling a way to increase profit? One rationale may stem from cost advantages like transaction cost reduction (Coase, 1960; Demsetz, 1968), economies of scale and scope (Paroush/Peles, 1981), savings in production costs or reduction in complexity costs (Fuerderer et al., 1998; Ringbeck et al., 1998). Obviously, increased sales and improving coordination through bundling can lead to cost reductions and better capacity utilization. Other rationales are the reduction of aggressiveness between duopolistic rivals (Carbajo et al., 1990) or the extension of the monopoly power (Burstein, 1960), as in the recent Microsoft case (Windows 95 with the Internet Explorer).

But the main rationale for bundling is to exploit the surplus of heterogeneous customers more effectively than can be done by using separate pricing (Stigler, 1963; Adams/Yellen, 1976). The consumer surplus is the difference between what the customer is willing to pay (his reservation price) and what he actually has to pay. Since the companies want to maximize profit, bundling makes sense only if the customers are heterogeneous with regard to the value they attach to the product and their willingness to pay for the individual products and the bundle. To illustrates the potential of bundling, let us assume that a restaurant offers wine and pizza. There are four consumers with the reservation prices (i.e., maximum willingnesses to pay) given in Table 2.

Table 2. Individual reservation prices for wine, pizza and the bundle

	Reservation prices (in DM)		
Consumer	Wine	Pizza	Bundle (wine and pizza)
1	9.0	1.5	10.5
2	8.0	5.0	13.0
3	4.5	8.5	13.0
4	2.5	9.0	11.5

For the sake of simplicity and without any loss of generalizability, we assume that costs are zero, i.e. we maximize sales revenue. In this example three forms of pricing are possible (Adams/Yellen, 1976; Simon, 1992):

- **Separate pricing:** Two units of both wine and pizza are sold and the profit is 8 x 2 + 8.5 x 2 = 33. This is the maximum attainable profit if we price and sell both products separately (no bundling).

- **Pure bundling:** The price for the bundle is set at 10.5 and consumers 1, 2, 3 and 4 buy the bundle (see Table 3 and Figure 2). The profit is now 10.5 x 4 = 42. This is 27.3 percent higher than the profit with separate pricing. Obviously, bundling increases profit substantially.

- **Mixed bundling**: Consumers 2 and 3 buy the bundle, consumer 1 buys only wine and consumer 4 only pizza. The profit is now 13 x 2 + 9 x 2 = 44. This form of price bundling is even more profitable. The profit is 33.3 percent higher than with "no bundling" and also 4.8 percent higher than with pure bundling. The results are presented in Table 3.

Table 3. Optimal prices, sales volume and profits for different pricing strategies

Pricing strategy	Optimal prices			Sales volume			Profit
	Wine	Pizza	Bundle	Wine	Pizza	Bundle	
Separate pricing	8.0	8.5	-	2	2	-	33.0
Pure bundling	-	-	10.5	-	-	4	42.0
Mixed bundling	9.0	9.0	13.0	1	1	2	44.0

Pure bundling reduces the number of segments in the market. With separate pricing we have four segments (buyers and non-buyers of the two products), with pure bundling we have only two segments (buyers and non-buyers of the bundle). This is illustrated in Figure 2.

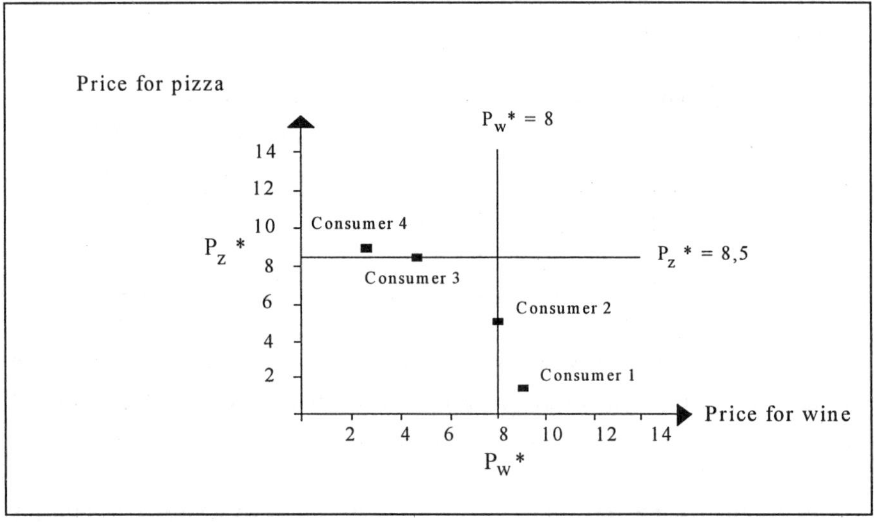

Fig. 2. Separate pricing

With pure bundling we offer the bundle for 10.5 (see Figure 3), we sell 4 bundles, and we and obtain a profit of 42. The profit increase is explained by the better exploitation of consumer surplus. The trick is that bundling transfers consumer surplus from one product to the other product.

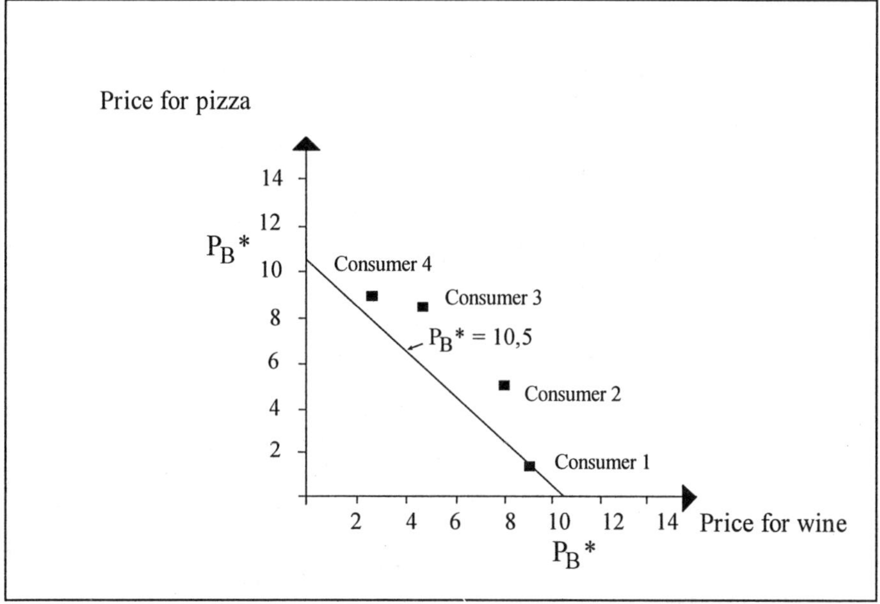

Fig. 3. Pure bundling

In our example, consumer 1 is willing to pay 9 for the wine. But with separate pricing we ask him to pay only 8. Thus, he retains a surplus of 1. On the other hand, his willingness to pay for pizza is 7 money units below the price of 8.5 for pizza alone.

Bundling means that we can add up the reservation prices (willingnesses to pay) and, thus, transfer excess consumer surplus from one product to another. In addition, the optimal bundle price is usually lower than the sum of separarate prices. Due to this combined effect, the bundle is usually bought by many customers - provided, of course, that the bundle price is set at the optimal level. A comparison between pure and mixed bundling forms reveals:

- With pure bundling consumers 1 and 2 buy wine in addition to pizza and consumers 3 und 4 buy pizza in addition to wine.

- With mixed bundling in addition to the bundle at a price of 13, we now offer wine and pizza separately, each at a price of 9. The new situation is shown in Figure 4.

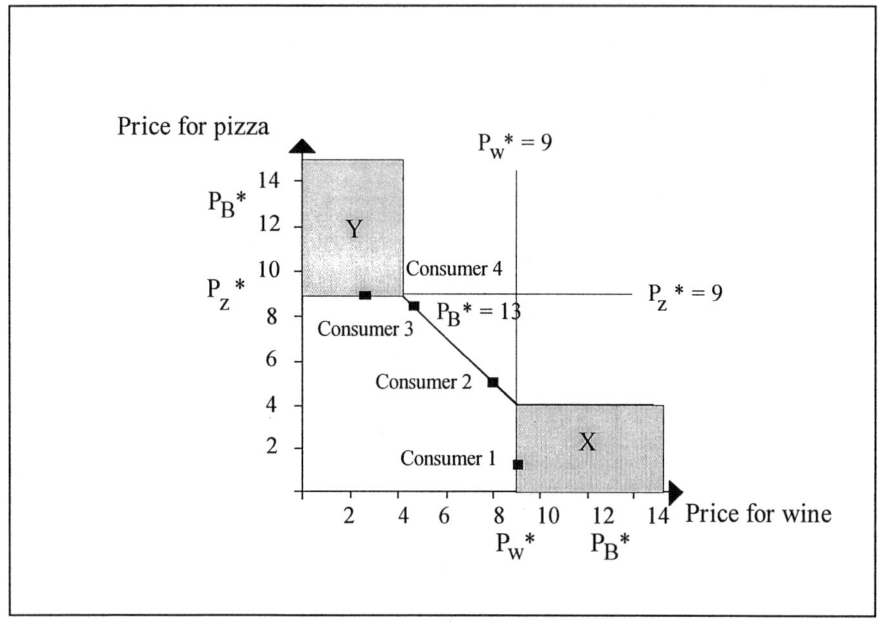

Fig. 4. Mixed bundling

Now consumer 1 is only buying wine and Consumer 4 is only buying pizza. Our profit rises to 44. Generally, consumers with reservation prices in the rectangles X and Y will become buyers of one of the products. Mixed bundling allows us to fine-tune the segmentation into bundle-buyers, single product-buyers and non-buyers very precisely. It is not possible, however, to give a general recommendation on which either method is optimal. This depends on the distribution of reservation prices. Roughly, the following advice can be given:

- If reservation prices are high for one product and low for the other product separate pricing tends to be optimal.
- If reservation prices are relatively high for both products pure bundling is recommmendable.
- If we have a combination of both customer groups, i.e. those with "extreme" preferences and those with “balanced” preferences, mixed bundling is probably the best pricing strategy.

In Table 4, we summarize and systemize the rationales for bundling based on the literature.

Table 4. Rationales for bundling, description, and the corresponding literature

Rationales for Bundling	**Description**	**Bundling Literature**
Extension of monopoly power	A firm with monopoly power in one market can use the leverage provided by this power to foreclose sales in, and thereby monopolize, a second market.	Burstein (1960) Warhit (1980) Whinston (1990)
Price discrimination	Bundling works as an implicit price discrimination tool, since it allows a seller to extract more consumer surplus from buyers.	Stigler (1968) Adams/Yellen (1976) Schmalensee (1984)
Reduction in complexity costs	Bundling (e.g. in the automobile industry) reduces the wide variety of options, which leads to a reduction in complexity costs.	Eppen et al. (1991) Anderson/Narus (1995)
Economies of scope and scale	The incentive for quantity discount arises from scale economies, whereas bundling yields scope economies.	Paroush/Peles (1981) Guiltinan (1987) Venkatesh/Mahajan (1993)
Reduction in transaction costs	Buyers avoid the transaction costs of contracting with several firms. Hence, they save time and information costs. Bundling reduces transaction costs.	Adams/Yellen (1976) Lawless (1991) Kohli/Park (1994)
Barrier to entry	Bundling can help to lock in customers, and hence deny them to competitors. Switching costs may be increased, acting as a barrier to new entrants.	Porter (1985) Lawless (1991)
Reduction of aggressiveness between duopolistic rivals	The authors consider a firm that has a monopoly over one good, while the other good is sold in competition with another firm. If the monopolists do not bundle, prices in the duopoly market are driven down to unit cost. But bundling differentiates the monopolist's product from that of its rivals. This induces less aggressiveness.	Carbajo et al. (1990)

5 Practical Applications of Bundling

5.1 Application of Bundling for Machine Tools

The problem in this case was to set the leasing rate for both the machine and the price for a maintenance contract, on a monthly basis. In addition, a decision had to be made whether the machine lease and the maintenance should be offered separately and/or as a bundle. The company also considered offering maintenance contracts for competitive products.

Table 5. Distribution of reservation prices for machine tools

Segment	Segment size (%)	Reservation prices (DM per month)			Remarks
		Leasing	Maintenance	Bundle	
1	12	1,250	990	2,310	
2	23	1,450	540	1,750	Companies with own service technicians
3	22	1,080	1,030	2,090	Very high usage
4	43	1,390	870	2,350	
Marginal Costs		550	470	1,020	

A study was commissioned to determine the reservation prices of actual and potential customers. Direct questioning was used, i.e. the customers were asked directly for the reservation prices. Based on this information, customers were clustered into four segments of different size. Table 5 provides this information. The reservation price for the bundle is sometimes lower (Segment 2) and sometimes higher (Segment 1 and 4) than the sum of the individual reservation prices. For Segment 3, both figures are similar.

Following the "no bundling"-strategy we obtain the following optimal values:

- Machine Price 1,390;
- Maintenance Price 870.

Segment 4 buys both machine and maintenance. Segment 2 leases the machine. Segments 1 and 3 buy the maintenance but not the machine. The profit index is 86,240. With pure bundling, the optimal bundle price is DM 2,090 and the profit index amounts to 82,390. Thus, pure bundling is in this case inferior to separate pricing. However, mixed bundling yields a considerably higher profit. In addition to the bundle for a price of DM 2,310, the machine tool lease is offered at

DM 1,450 and the price of the maintenance contract is DM 1,030. Total profit rises to 103,970, 20.6 percent above the profit obtained by the using separate pricing.

Table 6. Different pricing strategies, optimal prices, sales volume and profits

Pricing strategy	Optimal prices			Sales volume			Profit
	Leasing	Maintenance	Bundle	Leasing	Maintenance	Bundle	
Separate pricing	1,390	870	-	66	77	-	86,240
Pure bundling			2,090	-	-	77	82,390
Mixed bundling	1,450	1,030	2,310	23	22	55	103,970
Marginal costs	550	470	1,020				

5.2 Application of Bundling for Personal Computer Tools

A personal computer usually comprises three major components: the central processing unit (CPU), the monitor and the printer. In the following application for a large computer dealer, the reservation prices for the individual components and the bundle comprising of the three components were asked directly. Table 7 shows the value for twenty-five selected consumers.

Pure bundling is clearly inferior to both separate pricing and mixed bundling. Again mixed bundling yields the highest profit. The surplus in comparison to separate pricing is 7.7 percent (see Table 8).

Interestingly, CPU and printer are offered at optimal prices, while the price for the monitor can be set arbitrarily above DM 900, because this product will not be bought individually. In this case we recommend setting the price of the monitor at about DM 1,100 in order to demonstrate the advantage of the bundle price more strongly. It should be kept in mind that the color monitor is not actually bought individually at this price; it is only a pro forma offer. However, one should take care that this relatively high price does not lead to an unfavorable price image with customers.

Table 7. Individual reservation prices for personal computer components and bundle

Consumer	Reservation prices (in DM)			
	CPU	Color monitor	Printer	Bundle
1	1,400	400	650	2,150
2	1,700	650	600	2,700
3	1,600	750	700	2,800
4	2,200	600	680	3,400
5	1,900	600	750	3,100
6	2,000	800	800	3,400
7	2,500	800	600	3,500
8	1,800	500	650	2,600
9	1,700	600	800	2,700
10	3,000	400	1,000	4,400
11	2,000	600	700	3,000
12	1,500	900	600	2,700
13	1,100	500	600	2,000
14	1,700	700	800	3,000
15	1,500	650	450	2,200
16	2,000	700	700	2,900
17	1,600	200	400	2,000
18	1,300	700	700	2,500
19	1,700	700	600	2,600
20	1,400	700	600	2,300
21	1,450	500	500	2,100
22	1,500	800	800	3,100
23	1,500	600	800	2,700
24	1,500	600	500	2,100
25	1,700	600	500	2,700
Marginal costs	1,100	500	450	2,050

Table 8. Different pricing strategies and resulting profits

	Separate pricing			Pure price bundling	Mixed bundling			
	CPU	Color monitor	Printer		CPU	Color monitor	Printer	Bundle
Price	1,500	700	600	2,700	1,500	>900	600	3,000
Marginal costs	1,100	500	450	2,050	1,100	500	450	2,050
Sales	20	11	20	15	12	0	12	8
Profit	8,000	2,200	3,000	9,750	4,800	0	1,800	7,600
Total profit	13,200			9,750	14,200			

5.3 An Overview of Applications in the Bundling Context

Both applications in this article show that mixed bundling leads to the highest profit. An overview of the empirical studies in the bundling context (see Table 9) illustrates that mixed bundling always exploits profit potential better than separate pricing or pure bundling. The latter pricing strategy even leads to the lowest profits in all applications. This empirical finding may be explained by the fact that customers do not want to be forced to buy the bundle.

Table 9. Applications in the bundling context

Authors	Application	Number of bundle items	Profit increase relative to separate pricing (%)[a]	Optimal price bundle discount
Eppen et al. 1991	Software package	4	45	21
Simon 1992	Machine tools	2	20	16
	Computer options	3	8	6
Venkatesh/ Mahajan 1993	Season tickets	10	32	62
Fuerderer et al. 1994	Automobile option package	5	33	30
Simon 1995b	Automobile option package	2	20	21
Ansari et al. 1996	Season tickets	6	28	33
Huchzermeier/ Fuerderer 1997	Automobile option package	3	17	24
Wuebker 1998	Value meal	3	10	7

[a] In all applications, mixed bundling yields the highest profits and pure bundling the lowest profits.

Through mixed bundling, multi-product firms often obtain profit improvements in the range of 10 to 40 percent.

6 Implementation Issues in Bundling

6.1 Estimation and Optimization Aspects

Well-founded bundling requires the knowledge of customer specific reservation prices for both the individual products and the bundle. Several methods can be applied to measure these variables (see Wuebker, 1998).

The simplest method is to directly ask for the reservation prices. In many applications, this method is applied (Hanson/Martin, 1990; Venkatesh/Mahajan, 1993). Its reliability and validity may, however, be questionable because customers' attention is directly pointed to the price. This may induce an unrealistically high price-consciousness (see Simon, 1989, 27; Monroe 1990, 107). Another weakness of the direct method is described by Nessim/Dodge (1995, 72): "Buyers in direct responding may also attempt to quote artificially lower prices, since many of them perceive their role as conscientious buyers as that of helping to keep prices down." (see also Nagle, 1987, 280; Morton, 1989, 68; Monroe, 1990, 112). In industrial markets, the customers may have specific interests in naming a certain price to influence the supplier. Furthermore, by applying the direct method, price is viewed in isolation. But in reality, customers weigh the price against other product features, i.e., make a trade-off. The conjoint analysis addresses this trade-off issue in a systematic way.

Conjoint measurement can also be applied to obtain the information needed for bundling (Dolan/Simon, 1996; Wuebker/Mahajan, 1998; Wuebker, 1998). Bundles have to be included in the preference comparisons, and the reservation prices are indirectly calculated. The questionnaire design becomes rather complicated in this situation and the ability of the respondent to cope with this complexity has to be taken into consideration. In any case, the number of bundles to be investigated should be kept low.

A third method to measure reservation prices is expert judgment. However, in contrast to estimating the aggregate price response curve (Simon, 1989, 14), in the bundling context segment specific reservation prices have to be estimated. For the optimization, an aggregation at segment levels is always recommended. For small problems, an optimization "by hand" is feasible. This also applies if the individual reservation prices have been aggregated to yield a small number of segments.

Optimizing separate and bundle prices for a large number of respondents and components is a very complex problem. Hanson/Martin (1990) developed an optimization program which includes both the price and the composition of the bundle. This program compares separate pricing, pure and mixed bundling and selects the overall optimum. The following data have to be included in the model:

- Consumers' reservation price for all possible bundles.
- The size of the customer segments (formed by using cluster analysis).
- The costs of supplying customers in a specific segment with a certain bundle.

The program permits non-additive reservation prices and cost synergies for the bundle. Venkatesh/Mahajan (1993) and Fuerderer (1996) developed a stochastic optimization approach. The latter approach is described in the book (see Fuerderer et al., 1998).

6.2 Legal Problems and Restrictions

Bundling faces legal problems and restrictions. In a number of cases certain practices have been prohibited (see Table 10). In the USA block booking for films had to be abandoned after the decision in the Loew's case (see Phlips, 1989).

Table 10. Some antitrust cases in the bundling context

Source	Year	Company	Bundle of
224 U.S. 1	1912	A.B. Dick	mimeograph equipment with paper and ink
243 U.S. 502	1917	Motion Picture Patents	motion picture projector with films
298 U.S. 463	1936	IBM	tabulating machines with punch cards
332 U.S. 392	1947	International Salt	lixator with salt
356 U.S. 1	1958	Northern Pacific Railway	land with transport service
371 U.S. 38	1962	Loew's	blocks of movies
Court Decision in December	1997	Microsoft	Windows 95 with Internet Explorer

A well known case is IBM. The computer giant had to abandon the tying of tabulating machines and punch cards in 1936. In 1969, IBM also stopped bundling for computers in face of a threatened antitrust suit. Microsoft is in a similar situation in the late 90's (see Newsweek, November 3, 1997, pp. 42-44). Microsoft has a quasi-monopoly with the operating system Windows 95. Its world market share is over 80 percent (see Wuebker, 1998, 189). In 1995, when Microsoft introduced Windows 95, the firm bundled the Web-browser Internet Explorer with its market-dominant operating system. By applying this strategy, Microsoft pushed the market share of its Web-browser to 38 percent within two years (in June 1996, the market share of its Web-browser was just 7 percent). Competitors like Netscape and Sun Microsystems filed suit against Microsoft in the hope that the U.S Department of Justice will stop Microsoft's successful bundling strategy. Their argument is that Microsoft tries to monopolize the Web-browser market by abusing the market-dominant position of Windows 95. This practice ("extension of monopoly power") would violate the Sherman (§ 1, 2) and the Clayton Antitrust Act (§ 3) which prohibit bundling whenever the seller has sufficient economic power with respect to the tying product (e.g., Windows 95) to appreciably restrain free competition to the market for the tied product (e.g., Internet Explorer). The Assistant Attorney General Klein wants Microsoft to stop forcing manufacturers to include Internet Explorer and inform current customers how to wipe the Internet Explorer off the operating system. If Microsoft does not comply with these requests, Klein wants the court to fine it an unprecedented one million dollars a day. In this context, one crucial question will be: Is the component Internet Explorer a separate product with its own market (the argument of the petitioner) or just an additional feature or natural expansion of the operating system (the argument of Microsoft)? The latter has been permitted since the U.S. Department of Justice signed a consent agreement with Microsoft in 1995 (see Fortune, November 24, 1997, 127). Bill Gates hits the Microsoft case hard by asking: "Are we allowed to continue to innovate in products and in Windows itself?" (Newsweek, November 3, 1997, 44)

In Europe, the antitrust authorities also show a negative attitude toward bundling. In Germany, Effem GmbH, the leading manufacturer of pet food, had to give up its annual sales bonus system (a special form of bundling, see chapter 3). It was, however, only prohibited to grant sales bonuses on an annual basis; the quarterly sales bonus was considered legal (see Möschel, 1981 and Plinke, 1983). In a different case, the European Supreme Court ordered Hoffmann-La Roche to stop its loyalty bonus system for vitamins. Loyalty bonuses are also considered illegal in German jurisdiction (see WuW-Entscheidungssammlung OLG 2048). German antitrust agencies have looked into title combinations for newspapers (see WuW-Entscheidungssammlung OLG 1767). The soccer club 1.FC Köln was not allowed to sell tickets combining a more attractive and a less attractive soccer game en bloc (see WuW-Entscheidungssammlung BGH 2406). The legal restrictions apply mostly to companies with a dominant market position or companies which violate the so-called "Rabattgesetz". In these cases, it is strongly recommended that bundling be applied with caution and that the legal issues are settled before such a policy is implemented.

7 Summary and Conclusions

Bundling is a powerful method to better exploit profit potential and to maximize profits in a multi-product company. The heterogeneity of demand is reduced and customers' willingness to pay is used to the company's advantage.

Bundling can be applied in pure or mixed forms. In each individual case, it has to be carefully investigated which form is superior and how it compares to separate pricing for the individual products. There are no general or simple rules, but the optimal solution depends on the distribution of customers' willingness to pay. A valid measurement of these prices can be obtained by conjoint measurement. The profit increases attainable through bundling are likely to be in the range of 10 to 40 percent. Antitrust aspects should be considered, particularly if the company is in a dominant market position.

References

Adams, W. J. and Y. L. Yellen (1976). “Commodity bundling and the burden of monopoly.” Quarterly Journal of Economics, Vol. 90 (August), 475-498.

Anderson, J.C. and J. A. Narus (1995). “Capturing the value of supplementary services.” Harvard Business Review, Vol. 73 (January/February), 107-117.

Ansari, A., S. Siddarth and C. B. Weinberg (1996). “Pricing a bundle of products or services: the case of nonprofits.” Journal of Marketing Research, Vol. 33 (February), 86-93.

Blackstone, E. A. (1975). “Restrictive practices in the marketing of electrofax copying machines and supplies: The SCM corporation case.” Journal of Industrial Economics, Vol. 23 (March), 189-202.

Burstein, M. L. (1960). “The economics of tie-in sales.” Review of Economics and Statistics, Vol. 42 (February), 68-73.

Carbajo, J., D. de Meza and D. J. Seidmann (1990). “A strategic motivation for commodity bundling.” The Journal of Industrial Economics, Vol. 38 (March), 283-298.

Coase, R. H. (1960). “The problem of social costs.” Journal of Law and Economics, Vol. 22 (October), 1-44.

Cready, W. M. (1991). “Premium bundling.” Economic Inquiry, Vol. 29 (January), 173-179.

Cummings, F. J. and W.E. Ruther (1979). “The Northern Pacific case.” Journal of Law and Economics, Vol. 41 (October), 329-50.

Demsetz, H. (1968). “The cost of transacting.” Quarterly Journal of Economics, Vol. 82 (February), 33-53.

Drumwright, M.E. (1992). “A demonstration of anomalies in evaluations of bundling.” Marketing Letters, Vol. 3 (October), 311-321.

Dolan, R. J. and H. Simon (1996). Power pricing - How managing price transforms the bottom line. The Free Press, New York.

Eppen, G. D., W. A. Hanson and K. R. Martin (1991). "Bundling - new products, new markets, low risk." Sloan Management Review, Vol. 32 (Summer), 7-14.

Foster, I.R. (1991). "Cross-couponing as bundling." University Microfilms International, Ann Arbor.

Fuerderer, R., A. Huchzermeier and L. Schrage (1994). "Stochastic option bundling and bundle pricing." Working Paper 94-12, WHU Koblenz, Germany.

Fuerderer, R. (1996). Option and component bundling under demand risk. Gabler, Wiesbaden.

Fuerderer, R.; A. Huchzermeier (1997). "Optimale Preisbündelung unter Unsicherheit." Zeitschrift für Betriebswirtschaft, Vol. 62 (Ergänzungsheft 1), 17-133.

Guiltinan, J. P.(1987). "The price bundling of services: a normative framework." Journal of Marketing, Vol. 51 (April), 74-85.

Hanson, W.A. and K. R. Martin (1990). "Optimal bundle pricing." Management Science, Vol. 36 (February), 155-74.

Kohli, R. and H. Park (1994). "Coordinating buyer-seller transactions across multiple products." Management Science, Vol. 40 (September), 1145-1150.

Lawless, M. W. (1991). "Commodity bundling for competitive advantage: strategic implications." Journal of Management Studies, Vol. 28 (May), 267-280.

Liebowitz, S. J. (1983). "Tie-in sales and price discrimination." Economic Inquiry, Vol. 21 (July), 387-399.

Monroe, K. B. (1990). Pricing: making profitable decisions. 2nd edition, McGraw-Hill, New York.

Morton, J. (1989). "The economics of price." In: Seymour, D.T. (Hrsg.). The Pricing Decision - A Strategic Planner for Marketing Professionals. Probus Publishing Company, Chicago, 55-76.

Möschel, W. (1981). "Umsatzbonussysteme und der Mißbrauch marktbeherrschender Stellungen." Marketing Zeitschrift für Forschung und Praxis, Vol. 3 (November), 225-232.

Nagle, T. T. (1987). The strategy and tactics of pricing- a guide to profitable decision making. Prentice-Hall, Englewood Cliffs.

Nessim, H. and R. Dodge (1995). Pricing - policies and procedures. MacMillan Press, London.

Paroush, J. and Peles, Y.C. (1981). "A combined monopoly and optimal packaging." European Economic Review, Vol. 15 (March), 373-383.

Peterman, J. L. (1979). "The International Salt case." Journal of Law and Economics, Vol. 5 (October), 351-364.

Phlips, L. (1989). The economics of price discrimination. Cambridge University Press, Cambridge.

Plinke, W. (1983). "Periodenbezogene Mengenrabatte marktbeherrschender Unternehmen." Zeitschrift für betriebswirtschaftliche Forschung, Vol. 35 (March), 224-238.

Porter, M. (1985). Competitive advantage. The Free Press, New York et al..

Ringbeck, J., C.-S.Neumann and A. Cornet (1998). "Market-oriented complexity management using the micromarket management concept." Fuerderer, R., A. Herrmann and G. Wuebker (editors). Optimal bundling. Springer, New York et al..

Schmalensee, R. (1984). "Gaussian demand and commodity bundling." Journal of Business, Vol. 57 (January), 211-230.

Seidmann, D. J. (1991). "Bundling as a facilitating device: a reinterpretation of leverage theory." Economica, Vol. 58 (November), 491-499.

Simon, H. (1989). Price management. North-Holland, Amsterdam.

Simon, H. (1992). Preismanagement - Analyse, Strategie, Umsetzung. 2nd edition, Gabler, Wiesbaden.

Simon, H. (1995a). Preismanagement kompakt, Probleme und Methoden des modernen Pricing. Gabler, Wiesbaden

Simon, H. (1995b). "Pricing for the future - challenges and sophisticated solutions." Unpublished Presentation Paper, Chicago.

Simon, H., M. Faßnacht and G. Wuebker (1995). "Price bundling." Pricing Strategy & Practice, Vol. 3 (1), 34-44.

Sobel, R. (1981). IBM-Colossus in transition. The Free Press, New York.

Stigler, G.J. (1963). "United States vs. Loew's Inc.: A note on block booking." In: Kurland, P. (Hrsg.). The Supreme Court Review. University of Chicago Press, Chicago, 152-157.

Tacke, G. (1989). Nichtlineare Preisbildung: Theorie, Messung und Anwendung. Gabler, Wiesbaden.

Telser, L.G. (1965). "Abuse in trade practices: an economic analysis." Law and Contemporary Problems, Vol. 30 (Summer), 488-505.

Telser, L. G. (1979). "A theory of monopoly of complementary goods." Vol. 52 (April), 211-30.

Venkatesh, R. and Mahajan, V. (1993). “A probalisitic approach to pricing a bundle of products or services.” Journal of Marketing Research, Vol. 30 (November), 494-508.

Warhit, E. (1980). “The economics of tie-in sales.” Atlantic Economic Journal, Vol. 8 (December), 81-88.

Watson, T. J. Jr. (1990). Father, Son & Co., My life at IBM and beyond. Bantam Books, New York.

Whinston, M. D. (1990). “Tying, foreclosure, and exclusion.” American Economic Review, Vol. 80 (September), 837-859.

Wilson, R. B. (1993). “Nonlinear picing.” Oxford University Press, Oxford.

Wilson, L. O., A. M. Weiss and G. John (1990). “Unbundling of industrial systems.” Journal of Marketing Research, Vol. 23 (May), 123-138.

Wirtschaft und Wettbewerb, “Entscheidungssammlung”, Bundesgerichtshof (BGH 2406).

Wirtschaft und Wettbewerb, “Entscheidungssammlung (Kombinationstarif)”, Oberlandesgericht (OLG 1767).

Wirtschaft und Wettbewerb, “Entscheidungssammlung (Treuerabatte)”, Oberlandesgericht (OLG 2048).

Wuebker, G. (1998). Preisbündelung: Formen, Theorie, Messung und Umsetzung. Gabler, Wiesbaden.

Wuebker, G. and V. Mahajan (1998). “A conjoint analysis based procedure to measure reservation price and to optimally price product bundles.” Fuerderer, R., A. Herrmann and G. Wuebker (editors). Optimal bundling. Springer, New York et al., 157-175.

Part 2: Optimization Approaches

Optimal Price Bundling – Theory and Methods

Ralph Fuerderer[1]

[1] **Ralph Fuerderer**, Adam Opel AG, International Technical Development Center, Ruesselsheim, Germany.

1 The Challenge

In many production and service industries, *bundling* is the widespread practice of offering an array of products or services in a single package at one price.

A recent example in the German micro-computer branch is Vobis Microcomputer AG. Their rigorous bundling strategy is considered to be an important factor of their market leadership. Each customer can build his individual "Vobis bundle", which consists of tower, monitor, printer, and additional hard- and software. Through high component modularity, a customer can even set up his personal system configuration, including base module (processor, hard disc, RAM), keyboard, CD-ROM, as well as graphic and sound cards. Bundle discounts at Vobis generally range from 5 to 10 %, and relatively increase with the bundle price.

In the automobile industry, a car model is offered along with features such as air-conditioning, sun-roof, anti-lock brakes (ABS), navigation system, and so forth. These features may be part of the standard car offer (*standard options*), or must be purchased separately (*free-flow options*). As far as the competitive small, medium and compact size car markets are concerned, accessory and equipment sales can be an important source of profit for a manufacturer. Thus, it is vital to decide on the initial selection and pricing of free-flow options to be offered. Due to substantial product and process development lead times, this task has to be carried out at least several months before production actually starts. Currently, accurate forecasting of demand for particular car types or option combinations is extremely difficult.

The car producer can hedge his risk of not matching the individual preferences of the customers by providing a wide selection of free-flow options. However, in a global market environment, this product strategy is rather questionable from a development, production and marketing perspective. Engineering effort and variant-dependent manufacturing costs are prevalently impacted by the number and the design of option combinations a customer can purchase with his base car. Economies of scope exist among complementary options, e.g. a front door can be equiped with electrical mirror and window opener less costly, if it also has central door locks. However, these cost synergies can only be exploited if customers do select certain option combinations. On the sales side, a car vendor's service level and sales force effectiveness is largely depending on the number of options and option combinations.

These considerations favor option packaging under many economic circumstances. Consequently, "electric", "climate" or "sport" packages are used as strategic pricing instruments from the first day of a vehicle's life cycle on. As a means of stabilizing car sales volumes, special editions (e.g. Opel Astra "Cool") are successfully used by many world class manufacturers.

Conceptually, a bundle can be viewed as a new product, where the differentiation from existing products is achieved by deciding upon three crucial factors:

- the composition of the bundle
- the bundle price
- the presentation to the customer

However, bundling has some important advantages over designing a new single product. It starts with already available information on production and market conditions. This significantly reduces the high costs associated with a conventional product development and consequently decreases the risk of failure when a new product is brought to the market. Furthermore, bundling provides a relatively cheap opportunity to enhance differentiation from competitors and can even contribute to a positive public image. The potential profitability of bundling is mainly due to two sources. First, total costs of a bundle may be substantially lower than the sum of the costs for the individual items (cost sub-additivity).

Second, from the market side, bundling can be viewed as a tool to discriminate between customers with heterogeneous tastes, as well as a possibility to sell divisible goods at higher margins to consumers with strong specific tastes. Bundle pricing allows the vendor to move a step closer to the dream of every marketer, namely, value based pricing, that is, charge each customer his value received for his purchase.

Despite the variety of examples demonstrating bundling as an important product strategy, it has only received limited attention in the economic literature. A comprehensive state-of-the-art review will be provided in the next paragraph.

In section 2 we will focus on the three cornerstones of bundling mentioned above. Although an optimal strategy requires a simultaneous decision on composition, pricing, and presentation of a bundle, we will first investigate the specific aspects of each factor. We will show that it is advantageous when manufacturing, marketing and competition aspects are considered simultaneously, rather than exclusively limiting the effects of bundling to either cost savings, demand stimulation or product differentiation.

In section 3 we present the first mathematical programming approach to bundling by Hanson and Martin. It provides a practical method for a single firm to take a profit-maximizing decision on bundle designs and prices.

2 The State-Of-The-Art

Roughly speaking, the existing literature treats motivations and effects of bundling within frameworks of

- industrial economy
- competition legislation
- behavior
- economical analysis
- decision making

The earliest examinations in the economic literature on product bundling mainly assess antitrust and policy implications of bundling as a subtle price discrimination tool. In an influential paper, Stigler (1963) addresses "block-booking" and demonstrates that a movie distributor can raise profits by leasing only combined packages of movies instead of individual movies. The underlying idea is to prevent the customer from selecting "attractive" movies only and ignore inferior ones. In his analysis, Stigler introduces the concept of reservation prices. It depicts the maximum price a movie theater owner is willing to pay for a movie. He will purchase the movie, if his customer surplus, e.g., the difference between his personal reservation price and the actual price, is non-negative. Stigler concludes that this form of bundling is profitable, only if reservation prices are negatively correlated, while their aggregations (bundle reservation price) are similar for all customers. This enables the movie leasor to levy the excessive customer surplus through the bundle price, whereas the combination of lower and higher preferences for single movies in the package may lead to increased bundle prices. Many of the subsequent research is operating with this analytic concept.

In a related line of research, Burstein (1960ab), Telser (1965) and Stigler (1968) investigate *tie-in sales*, which require customers to purchase a commodity (tied good) with a focal product (tying good). In one example, IBM tried to tie the sales of punch cards to tabulating machine sales. There are two aspects to consider: first, one could interpret this system as an effort of IBM to expand the tabulator monopoly to punch cards. Telser (1965), however, has the opinion that punch card usage is a measure of machine utilization, and thus, machine utility. Due to the freedom to charge a high mark-up on punch cards within a bundle, IBM succeeded in exploiting their customers utility according to the intensity of tabulating machine usage. This would not have been possible without tying in card sales, since each customer paid the same price for the tabulating machines. Tie-in sales extract consumer surplus more effectively than single pricing, which is not appropriate to quantify customers utility. A similar example is reported by Blackstone (1975), where Xerox only leased its copy machines to its customers and based prices on the number of processed copies. Firstly after 1962, they additionally offered a sales option which, however, was so unattractive that most customers held on to the leasing option.

The important essay of Adams and Yellen (1976) introduces a two-dimensional graphical framework for analyzing the effect of bundling as a price discrimination tool. They examine a multiproduct monopolist with two products, independent and additive consumer valuations and linear demand for these goods.

They consider additive variable costs and no fixed costs associated with each commodity. In addition, they assume that a customer will either purchase one unit of a commodity, or will not purchase it at all (zero marginal utility of a second unit). Comparing unbundled sales to *pure bundling* (offering only the complete bundle) and *mixed bundling* (offering both the bundle and subsets of the bundle), Adams and Yellen show by examples, that each pricing strategy can be advantageous. The two major factors determining the profitability of either strategy are the level of cost and the distribution of customers in the reservation price space. A strongly negative correlation of reservation prices, for instance, is shown to favor bundling strategies as opposed to single pricing. For a symmetric demand distribution and a relatively high cost level, a single pricing strategy may be preferred. Moreover, Adams and Yellen note that the sales of physical products in different container sizes (detergents, drinks and food) can also be categorized as a form of mixed bundling. In modern literature, this practice is rather attributed to non-linear pricing (for details see Tacke (1989) and Simon (1992)). Although non-linear pricing also aims at capitalizing on consumer price readiness, its theoretical justification stems from *Gossen's Principle* (Gossen (1854)) of decreasing product utility with increasing quantity.

In a similar line of research, Cready (1991) addresses profitability conditions for *premium bundling* (the bundle price is higher than the sum of component prices). In case of overall positively correlated reservation prices and a strong negative correlation among consumers with relatively low reservation prices, premium bundling can be more profitable than unbundled sales, pure or mixed bundling. The major practical problem, of course, is how to prevent customers from self-bundling the bundle components. Although he mentions couponing and rebates as potential tools to employ premium bundling, realizations of this practice, as for collectible items (stamps, coins), seem to be rare.

Dansby and Conrad (1984) drop the assumption of additive reservation prices. They examine the case, where a bundle may either contain an unwanted component (value reducing) or provide additional value beyond the aggregated values of the individual items (value enhancing). They conclude that this diversity in consumers' bundle preferences can be an additional incentive for firms to bundle, even if the firm has no monopoly power.

Schmalensee (1984) for the first time conducts a more formal analysis and develops numerical criteria on which a bundling strategy turns out to be more profitable. Thereby, he assumes that buyers' reservation prices follow a bivariate normal distribution. Stating the profit function and the according extremal problem, he shows that there is no explicit representation of an optimal price. Thus, he obtains profitability conditions in either bundling case by numerical analysis. He extends the result of Adams and Yellen for pure bundling and shows that a negative correlation in consumers' reservation prices is no necessary condition for pure bundling being more profitable than unbundled sales. Actually, pure bundling is shown to be always more profitable, if reservation prices have a positive correlation and the cost level is low enough. Furthermore he shows that mixed bundling is not more profitable than pure bundling in the case of perfect positive correlation of reservation prices. As in the pure bundling case he suggests that mixed bundling is more profitable than unbundled sales, if the cost level is low enough, but he claims that his analysis of an optimal mixed bundling strategy is not complete.

Extending the above result, Salinger (1995) suggests through graphical analysis that in the case of relatively high costs, positively correlated reservation prices may increase the incentive to bundle if economies of scope prevail.

Eppen, Hanson and Martin (1991) present seven strategic guidelines for the successful implementation of a bundling strategy, which partly accrue from a qualitative interpretation of the Hanson/Martin model we will present in section 1.4. They particularly address the opportunity of bundling to expand demand ("aggregation bundling", "trade-up bundling" and "loyalty bundling"). Other strategic goals supported by bundling are expansion of monopoly power to competitive markets (Warhit, 1980; Palfrey, 1983) as well as increase of product enhancement and possibilities of product differentiation from competitors (Porter, 1985).

A number of recent studies address the advantages of unbundling commodities in competitive markets, which seems to become more popular over time (Jackson, 1985). Whereas Schmalensee (1982) states that a monopolist will never choose to bundle its indivisible goods with the goods of a perfectly competitive industry, if the goods are unrelated in production and consumption, Carbajo, De Meza and

Seidman (1990) show that the monopolist may well bundle with the product of an imperfectly competitive industry. They explain their findings with the strategic role of bundling in the presence of imperfect competition as a mean to induce rivals to compete less aggressively. Considering industrial systems, Porter (1985) identifies two managerial decisions that system suppliers are facing. A system supplier may maintain its current strategy and try to strengthen its current market position by using advanced technology to offer customers higher system benefits. The second option is to unbundle complete systems and sell system components separately. This enables the supplier to possibly withdraw from the market for some system components and "outsource" these products.

Wilson, Weiss and John (1990) provide some market conditions, which favor unbundling, as potentially higher margins for unbundled systems which stem from a reduced price elasticity of the single components may prevail. Besides demand expansion issues, they also mention increasing integration and modularity: this raises the risk of pure bundling strategies, because customers are able to obtain desired components by simply mixing and matching components from different suppliers. Guiltinan (1987) compares mixed bundling strategies and unbundled sales with complementary consumer services. He particularly stresses the necessity of clear strategic marketing objectives and a thorough analysis of the specific bundling program.

Other marketing investigations aim at the customers' perception of bundled products. It is well known (Gaeth et al. (1991)) that in particular customers with a low information level are willing to pay a higher price for the bundle than the prices for the individual items would suggest. It is astonishing though and must be considered a bad sales practice that in car purchases, for instance, customers are induced to buy individual car options which are a subset of an offered package, at a higher price than they would be charged for the entire package. In this framework, Yadav and Monroe (1993) examine the transaction value of a bundle with a focus on customers perception of savings in a bundle price. They promote offering two small savings (on the purchase of all bundle items separately and on the bundle purchase) instead of offering only one large saving on the bundle. They stress however, that results seems to depend on the semantic format of the specific presentation method. Nagle (1987) generally suggests that bundled items at one price are more likely to induce a customer to purchase them, than separately offered and priced products. A comprehensive treatment of behavioral aspects is presented in part 2.

In the preceding essays, only little insight was gained on the optimal composition of a bundle and the issue of optimal pricing. Most recent contributions consider general product line design in the context of conjoint analysis. Zufryden (1977,82) was the first to formulate a share-maximizing product line problem as an integer program, assuming deterministic preferences. He suggests to select the product line directly from idiosyncratic part worths data obtained by conjoint analysis (for an exhaustive review of commercial applications of conjoint analysis see Cattin and Wittink (1989)). A similar line of research follow Dobson and Kalish (1993) or Bauer, Herrmann and Mengen (1994).

Kohli and Sukumar (1990) present a 0-1 integer formulation how to structure product lines maximizing share, buyer's utility and seller's profit. Since the particular problems are NP-hard, they also propose a dynamic programming heuristic which extends the result of Kohli and Krishnamurti (1987) for choosing a share maximizing single item. Other contributions by McBride and Zufryden (1988), and Green and Krieger (1992) consider a finite reference set of candidate items from which they select the product line. Dobson and Kalish (1988) measure consumer's utility in reservation prices and consumer's choice behavior by the obtained surplus. They explicitly consider fix and variable production costs and propose heuristics to solve the resulting non-linear problem to maximize profit. In general, most deterministic mixed integer programs are flexible and can be well adapted to the institutional structure of the problem. On the other hand, this approach is frequently computationally intractable and objective functions are non-concave.

Stochastic choice models mainly focus on consumer segmentation and utility measurement (Ogawa, 1987), whereas many contributions explored the aggregate utility structure as linear interaction (Green, 1984), quadratic (Louviere and Woodworth, 1983), logit (McFadden, 1974), probit (Daganzo, 1979), maximum score (Manski, 1975), or Generalized Extreme Value (McFadden, 1980). Particularly widespread is the multinomial logit model (McFadden, 1986) and has been applied by many authors to product line design and pricing (e.g. Kamakura and Russell, 1989; Allenby and Rossi, 1991). In a recent contribution, Hanson and Martin (1994) present an interesting approach how to optimize a non-concave profit function using a multinomial logit model. They point out that a lack of at least quasi-concavity of the objective may cause any solution method to terminate at a local optimum. To avoid this, they apply a path-following procedure (Garcia and Zangwill, 1981) which defines a "path" between the original non-concave problem and a concave problem with a higher degree of randomness in consumer's choice behavior. Starting with the "easy" modified problem, they use the optimal price vector of a preceding step as a starting point for a subsequent less concave problem. Thus, they drag along a closed path with constant increment for the homotopy defining parameters and end up solving the original problem after a finite number of steps. Although this method does not guarantee to find the global optimum of a function, several optimization runs with different parameter sets increase confidence in the quality of the obtained result. We will follow the stochastic approach in detail in section 4.

In the next section, we take a look at the cornerstones of a bundling strategy as presented in section 1.

3 The Dynamics

3.1 The Composition of the Bundle

A bundle consists of two or more products or services which may have complementary, substitutive or independent character.

From a cost perspective, bundling should be encouraged among components that have high setup costs and benefit from economies of scope. In car manufacturing, most of the door options as central door locks, power mirrors windows, burglar alarm and radio use an expensive door plug which connects each door wire with the corresponding switches and the power source. This plug must be installed, if one of the mentioned electrical options is required in the car, and it can subsequently be used by any other of them. Thus, marginal wiring harness costs for the second or third door option only consist of the material costs for the additional wires and increased cable installation times.

In both, producing and service industry, bundling two products aims at utilizing economies of scale for at least one of them. Including a jeep safari in a Kenia travel package will - if reasonably priced - clearly increase the safari company's customer base and positively impact their operational costs.

On the demand side, the main idea of offering a collection of components as opposed to sell them individually is similar to the idea of price promotions: a bundle will increase profit, if the discount is outweighed by the additional demand for the package. Market expansion is a key factor of bundling (Eppen, Hanson and Martin (1991)). To increase demand, it is necessary to understand which elements of the product line are valued by the consumers. In the safari example, reducing the traveller's transaction costs and improving the travel package's attractiveness may even result in higher overall demand for the entire package.

The choice of the "ingredients" of a bundle for instance enables the marketer to design a bundle for various customer segments which have a large intersection of valued single items: for three items A, B and C we may have a segment which strongly prefers items A and C, another one values B and C. By attractively pricing bundle ABC, we may achieve that both segments will purchase it. Especially in the car industry, hierarchical bundles are used to create a trade-up effect: the car's trim levels usually start with a basic car equiped with few standard features. Subsequent trim levels are defined by successively adding equipment to the last trim level. The idea is to persuade a customer to "trade up" to the next trim level. Eppen, Hanson and Martin (1991) suggest that the number of trade-up bundles generally is determined by demand: strong demand favors few offerings and more of a pure bundling approach, weak demand favors more alternatives with smaller gaps between bundles.

An important aspect of bundling in service industries is to make it unattractive for the customer to switch brands by reducing customers transaction costs and raising switching costs. A variety of American banks already use bundled accounts which comprise credit cards, checking accounts, certificates of deposit, and loans. The convenience of customers of doing all their banking with one institution is one of the biggest selling points used to push the bundled accounts. As a scale effect, bundled accounts reduce the overhead and support costs of taking in new deposits and making loans. A similar phenomenon can be observed in cellular industry. AT&T, for instance, bundles long distance calls with domestic offerings at a discounted price, and thus significantly reduces consumer trial and switching.

A crucial point of a bundling strategy is, whether the bundle concept succeeds in clearly defining the product. A consumer, who is not familiar with a product or service, may need an overall idea of the intention of a particular bundle (e.g. a safety package in a car, consisting of dual airbags, ABS, traction control and a reinforced passenger cavity). Even, if a customer has no experience with traction control, he may feel safe to order it in the company of some well-known safety features.

3.2 The Bundle Price

In this paragraph we want to describe the basic assumptions and mechanisms, which underlie optimal pricing methods for a particular bundling strategy.

A widely used idea to describe demand is the concept of a customer's reservation price already described. In a formal derivation of reservation prices, Kalish and Nelson (1988) show that a consumer's reservation price for a product is equal to the total utility contribution of the product divided by the marginal utility of money. We will assume that a customer purchases either one unit of a certain product or does not purchase anything. Thus, if a customer has several offers, he will choose the one which maximizes his personal surplus, a synonym for utility in this context.

Mostly in practice and theory, demand is not considered at the individual consumer level. Similar consumer preferences are accumulated to an aggregated reservation price of an entire customer segment. This deterministic approach assumes that the variance of consumers reservation prices within a segment is very small, and is convenient to understand the “dynamics” of bundling.

In practice, one can observe that reservation prices are sub-additive which means that the reservation price for an entire component bundle is not greater than the sum of reservation prices for the single items. Of course, there exist exceptions from this rule. If a package of products increases their overall utility, as observed for collectible items, reservation price may be super-additive for these components. According to Lewbel (1985), a customer's reservation price will be sub-additive, if he perceives the bundled goods as substitutes and super-additive, if the goods are complementary. In general, a distinction between substitutability and complementarity is difficult. As Dansby and Conrad (1984) point out, the utility of a bundle can also be less than the sum of the individual items' reservation prices when certain restrictions are imposed on the use of the bundle. Our Kenia travel package, for instance, may only be available with some restrictions on the travel time.

Under the objective to maximize profit by optimally extracting consumer surplus, sub-additive reservation prices imply sub-additive product pricing. If a customer purchases a bundle of components, he usually gets a discount on the price of the single items. If costs are strictly additive, bundling reduces the profit obtained by those customers, who are willing to purchase all components individually. However, bundling may create additional demand, and thus profit, from consumers, who did not purchase any component or purchased only some components of the bundle.

Profit increases and gains of market shares due to bundling may be substantial. According to Sobel (1981), in the IBM tie-in sale case (which relates to pure bundling, see section 1.2), punch card sales - as a direct consequence of tabulator sales - contributed by almost 40% to IBM's overall annual revenues.

3.3 The Presentation of Bundling

We shall mainly concentrate on the concepts of unbundled sales, pure and mixed bundling, as already presented.

A recent example of pure bundling is the ticket sales practice for the soccer world championships 1994 in the US. It was not possible to purchase first- round tickets for one particular event such as the opening match of Germany and Bolivia in Chicago. Tickets were only offered en bloc, comprising several matches for one location which lead to decent sales volumes even for the potentially unattractive matches (which did not prevent brokers to unbundle ticket packages and profit, however). Another example is the block-booking case discussed in section 1.2 (Stigler, 1963, 1968).

In service industry, maintenance contracts for computer hardware frequently consist of a variety of services which are not offered individually to the customer. A special case of pure bundling are tie-in sales, where the buyer of a tying good agrees to purchase any tied goods only from the distributor of the main product. The tying good mostly is a machine (computer, copier), the tied good is a required resource to run the machine (data discs, copy paper).

Examples of mixed bundling can be encountered in almost every business branch. It is particularly common in service industry (Guiltinan, 1987). Similar to our Lufthansa example, banks offer credit cards at no annual fee and free traveler's checks for “special” customers. Insurances sell bundles consisting of liability plus accident insurance. Hotels are offering weekend packages that combine lodging and meals at special rates, restaurants attract their guests with attractively priced lunch menus (starter, main course, desert and drink). In health clubs, a customer can choose between individual programs (aerobic classes, weight room, sauna), packages combining several of these activities or a universal membership. Airlines offer frequent flyer programs, where a certain amount of accumulated flight miles means free round-trips for the passenger. The latter idea is closely related to couponing which can also be viewed as a form of mixed bundling. Discounts on a mixed bundle can occur in two forms (Guiltinan, 1987): in “mixed-joint form”, there exists an explicit price for the bundle, in “mixed-leader form”, the consumer obtains a discount on one product, if he purchases another product at a regular price.

A very important aspect of bundling in a production framework is deproliferation (complexity reduction). In particular, pure bundling can ex ante reduce product and process complexity to a large amount. This feature is not so obvious for mixed bundling. As long as all products in a bundle can still be purchased as single items, the number of possible product combinations is not reduced. However, achieving a large penetration for a mixed bundle by attractive prices for the bundle and increased prices for particular individual items, is practically equivalent to pure bundling. On the demand side, pure bundling obviously not only reduces product complexity, but also limits product variety for the customer.

In general, as Schmalensee (1984) points out, pure bundling reduces effective buyer heterogeneity by aiming at aggregated reservation prices. The strength of unbundled sales is the ability to collect high prices from consumers with extreme tastes. Mixed bundling combines the features of pure bundling and unbundled sales. In general, the decision on the bundling strategy is a tradeoff, determined by cost and utility levels, as well as correlation of demand for the considered components.

McAfee, McMillan and Whinston (1989) show that for independently distributed preferences, (mixed) bundling always dominates unbundled sales. However, even under the fairly strong assumption of normally distributed demand (Schmalensee (1984)), there is no general rule, whether bundling is more profitable than unbundled sales. We want to illustrate the simple case of a two dimensional price space. Demand is perfectly transparent and represented by means of customer segments and fixed reservation prices for each segment. The following example shows, that either mixed bundling or unbundled sales can be preferable for a given (discrete) distribution of reservation prices.

Example 1:

We consider four customer segments with symmetrically distributed and negatively correlated reservation prices for A (10,95), B (40,80), C (80,40), and D (95,10). Costs for both components are set to 20 and 40 for the bundle. Figure 1 shows that under each strategy the reservation price space is split into several regions, four for unbundled sales and mixed bundling, two for pure bundling.

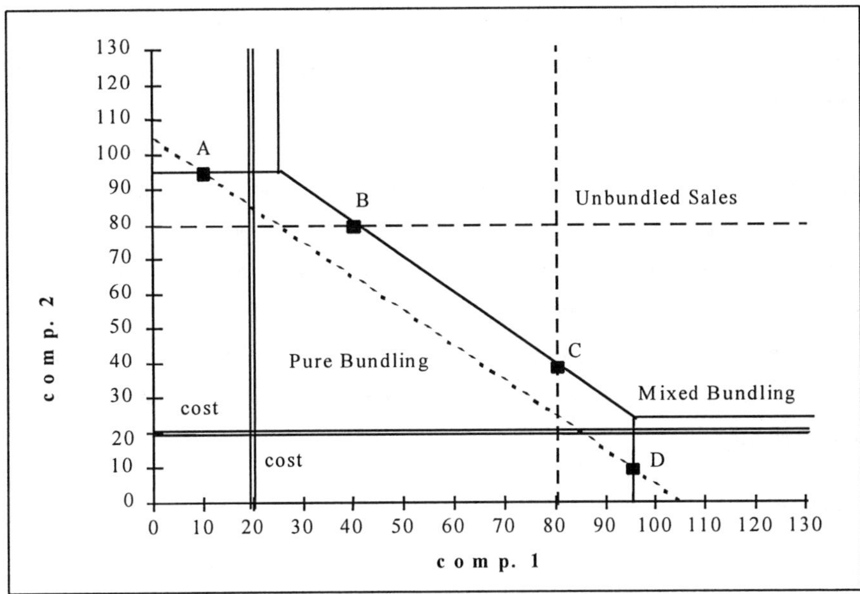

Fig. 1. Negative correlation of reservation prices

For each pricing strategy we state the optimal individual prices p_1, p_2, the optimal bundle price p_B and the resulting overall profit below.

Table 1. Optimal individual prices

Pricing Strategy	p_1	p_2	p_B	Profit
Unbundled sales	80	80	--	240
Pure bundling	--	--	105	260
Mixed bundling	95	95	120	310

Demand for the bundle is elastic, since its reservation prices only vary between 105 and 120 among all four customer segments. However, demand for the individual components varies between 10 (below cost) and 95. It seems reasonable to prefer an unbundled sales strategy, if we only deal with extreme tastes as for segments A and D, because consumer surplus can be completely extracted and profit margins are high. A complete extraction of consumer surplus in any bundling strategy would mean to lower profit margins, because the reservation prices for the less preferred items are below costs. Adding consumer segments B and C changes the situation. To include segments B and C in unbundled sales, we have to decrease price from 95 to 80 for each component which lowers margins. Including these segments in a pure bundling strategy leaves the optimal price of 105 unchanged, since the reservation prices r(B), r(C) are greater than 105. However, consumer surplus can not be completely extracted anymore (s(A)=s(D)=0, s(B)=s(C)=15). Through mixed bundling, surplus of B and C can be fully extracted by the bundle, as well as the surplus of A and D by their preferred single components. Since the margin of the bundle is not much greater than for the single items, mixed bundling turns out to be most profitable in this example.

Example 2:

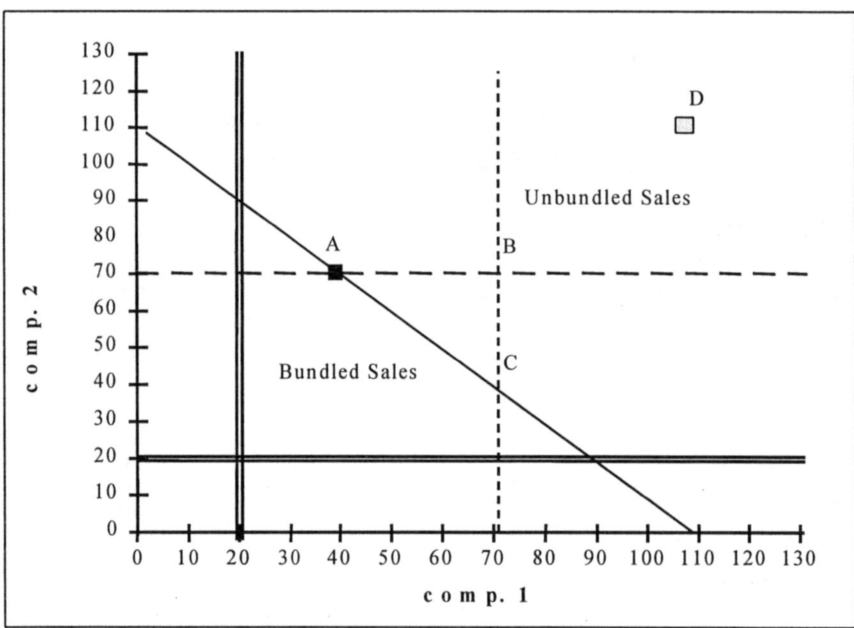

Fig. 2. Positive correlation of reservation prices

We now consider four customer segments with symmetrically distributed and positively correlated reservation prices, in particular for A (40,70), B (70,70), C (110,110) and D (70,40) (see Figure 1.2). Optimal prices and resulting profit for costs (20,20) are now:

Table 2. Optimal prices and resulting profit

Pricing Strategy	p_1	p_2	p_B	Profit
Unbundled sales	70	70	--	300
Pure bundling	--	--	110	280
Mixed bundling	>70	>70	110	280

Demand for the bundle is inelastic now, reservation prices vary between 110 and 220 among the segments, and range from 40 to 110 for the single items. Considering only customer segments A and D, we would always prefer bundled sales as opposed to unbundled sales. Although consumer surplus can be completely extracted under both strategies, tying the less preferred single item in the bundle raises margins. Adding segments B and C to the demand scenario does not change the optimal price for the bundled sales: gaps between the four bundle reservation prices are so large that no segment can be excluded without lowering profit. Consumer surplus cannot be sufficiently extracted anymore (s(A)=s(D)=0, s(B)=30, s(C)=110). In the unbundled case, optimal prices do not change either, but the surplus of customer segment B can be fully extracted (s(B)=0) and the surplus of segment C is lower than in the bundled case (s(C)=80). This more than compensates the slightly higher margins in bundling. We also observe that in this case mixed bundling yields the same profit as pure bundling: individual items under mixed bundling have to be priced higher than 70 in order to exclude segments A and D from the less profitable single component purchase (a single component price of less than 70 would include them in both cases single purchase and bundle purchase, and they would choose the alternative, which maximizes their surplus). But then, on the other hand, surplus for each segment will always be maximum, if the entire bundle is purchased, thus no single component will be sold. The optimal pricing strategies for pure and mixed bundling are identical in this example.

4 The Decision Models

4.1 Enumerative Approaches

In this line of procedures, all or a subset of all possible bundle configurations are enumerated, and a "test" price is assigned to every test bundle. Each of these scenarios can now be evaluated regarding their profitability.

Not surprisingly, enumerative models are widely used in practice. Potential reasons are:

- Elaborated methods are rather unknown.
- In decision making situtations of little complexity, employment of sophisticated methods is not considered cost efficient.
- Digital data processing allows for straightforward implementation of random simulation (Monte-Carlo) or directed simulation methods.

Enumeration and evaluation of alternatives provides an opportunity to support the economic decision making process. However, the disadvantages of this approach are obvious:

- Even for problems of limited scope, a total enumeration of bundle configurations is rather infeasible. For three products, there exist 116 mixed bundle configurations. For four products, more than 44.000 alternatives need to be priced and examined.
- In most cases, partial enumeration leads to suboptimal results with unpredictable quality of the solution.
- The character of price as a continuous variable adds another source of imprecision and unpredictability to the discretized procedure.

In the next paragraph, we will present the decision model of Hanson and Martin (1991). This mathematical programming approach for the first time allowed to determine the profit maximizing bundle configurations and prices, without explicitely considering the full range of feasible problem solutions.

4.2 The Hanson and Martin Model

The Hanson / Martin model is based on the following sequence of events. A profit maximizing monopolist company establishes a price list for all possible bundles of components. Customers examine the price list and make purchases which maximize their consumer surplus. The model developed is from the firm's perspective, and the objective function represents the firm's profit. The behavior of the customer as maximizers of their surplus is treated in the model as constraints on the firm's objective function.

Problem Parameters:

S = $\{1,\ldots,n\}$ is the index set of component items.

h = index of components, $h=1,\ldots,n$.

i = index of component bundles, $i=1,\ldots,L=2^n-1$.

k = index of customer segments, $k=1,\ldots,m$.

R_{ki} = reservation price of customer segment k for bundle i, $R_{k\{h\}}$ denotes the reservation price of the bundle which consists of only component $\{h\}$.

c_{ki} = cost of supplying one customer of segment k with bundle i.

N_k = number of customers in segment k.

$B(i)$ = the set of components which define bundle i.

$I \subseteq \{1,\ldots,L\}$ is an index set of bundles.

Decision Variables:

p_i = price assigned to bundle i.

θ_{ki} = 1 if customer k selects bundle i, 0 otherwise.

s_k = consumer surplus obtained by customer segment k.

Auxiliary Variables:

s_{ki} = consumer surplus of a customer in segment k if that customer selects bundle i.

z_{ki} = the marginal revenue generated from a customer in segment k if that customer selects bundle i.

p_{ki} = the price a customer in segment k pays for selecting bundle i.

The model can be viewed as a generalization of the approach of Stigler, as well as Adams and Yellen. The core assumptions are as follows:

Model Assumptions:

1. The benefit of a duplicate component is zero, and component resale is not possible.
2. A single profit maximizing firm determines the price of every bundle.
3. Given these prices, every customer maximizes his/her consumer surplus. Consumer surplus equals the reservation price minus product price. purchasing nothing yields zero consumer surplus.
4. All customer segments face the same prices.
5. The reservation prices of all customer segments are known for all bundles.
6. There is free disposal of unwanted components. Thus $B(i) \subseteq B(j) \subseteq S$ implies that the reservation price of bundle j is greater or equal to the reservation price of bundle i.
7. Marginal costs of a bundle are subadditive. This means that given any bundle B(i) and an index set I, $B(i) \subseteq \cup_{j \in I} B(j)$ implies the marginal cost of B(i) cannot exceed the sum of the marginal costs for bundles B(j), $j \in I$.
8. Customers have zero assembly transaction costs for creating bundles from separately offered subbundles.

Unlike Stigler or Adams and Yellen, there are no restrictions on the number of components offered, the structure of the reservation prices (i.e., strictly additive), or customer segment sizes. Free disposal of unwanted components, assumption 6, is a restriction on allowable reservation prices which implies that no customer segment would strictly prefer not to have a component included for free. While not holding for all products or components (for example, air conditioning is disliked by some automobile buyers), it is important for the algorithm we present.

The same is true for the restrictions on cost contained in assumptions 7 and 8. Subadditive bundle costs are a form of (weak) economies of scope, where the presence of one component in a bundle does not raise the cost of including other components, and may make it cheaper. Zero assembly transaction costs for creating bundles, assumption 8, is an insistence that bundles are well specified. If costly assembly must be performed on components *A* and *B* to achieve bundle *AB*, then '*AB*' really contains the additional component *C* representing assembly. However, the firm may well find it profitable to only offer the items *A*, *B*, and *ABC* in its product line.

By assumption 3, each customer maximizes his or her surplus. Thus, a customer will buy every bundle which yields a positive surplus. However, the union of all the components in each bundle purchased also constitutes a bundle. For example, if there are two components, *A* and *B*, and each component has positive surplus, the consumer would purchase both, *A* and *B*. For model formulation purposes, it is much easier to assume that the customer buys only one bundle. Thus we want to show that the customer is never better off buying more than one bundle. In order for this to be true prices must be subadditive. Indeed, if the price of *AB* were greater than the price of *A* plus the price of *B* the customer would prefer to buy *A* and *B*, over buying *AB* (since there are zero assembly costs by assumption 8). First it is important to note that given assumptions 1-8, a firm can always maximize profits by using subadditive pricing. Moreover it is shown that one may assume that the customers purchase at most one bundle. The proofs are omitted here and can be found in the original article.

Obviously it is impractical to package or list prices for every $(2^n - 1)$ bundle when n is large. However, this does not cause an implementation problem. In the bundle pricing model given below, if there are *M* customer segments then at most *M* distinct bundles are part of an optimal solution. Only the optimal bundles need be offered as long as the optimization model considers every bundle in determining which *M* bundles to offer. Note the crucial distinction between *offer* and *consider*. A bundle is considered in the model if that bundle can be a part of the optimal solution. A bundle is offered when it is part of an optimal solution. Therefore, when offered the *M* bundles from the optimal model solution the consumers behave exactly as they would given the complete list of optimal prices. Indeed, a key result of the bundle pricing solution is the choice of which products to offer out of the vast number of possibilities. Given assumptions 1 through 8, the constrained optimization model for maximizing the firm's profit is given in (1) to (11).

Objective function

$$\max \sum_{k=1}^{M}\sum_{i=1}^{L} N_k z_{ki}, \tag{1}$$

Consumer surplus

$$s_k \geq R_{ki} - p_i, \quad i = 1,\ldots,L,\ k = 1,\ldots,M, \tag{2}$$

Price subadditivity

$$p_i \leq \sum_{j\in I} p_j, \quad s.t. \bigcup_{j\in I} B(j) = B(i),\ i = 1,\ldots,L, \tag{3}$$

Single price schedule

$$p_{ki} \geq p_i - (\max_{\substack{k=1,\ldots,M \\ i=1,\ldots,L}} \{R_{ki}\})(1-\theta_{ki}), \quad i = 1,\ldots,L,\ k = 1,\ldots,M, \tag{4}$$

$$p_{ki} \leq p_i, \quad i = 1,\ldots,L,\ k = 1,\ldots,M, \tag{5}$$

Tightening constraints

$$s_k \geq \sum_{i=1}^{L} (R_{ki}\theta_{ji} - p_{ji}), \qquad k = 1,\ldots,M,\ j = 1,\ldots,M, \tag{6}$$

Single purchase

$$z_{ki} = p_{ki} - c_{ki}\theta_{ki}, \qquad i = 1,\ldots,L,\ k = 1,\ldots,M, \tag{7}$$

$$s_{ki} = R_{ki}\theta_{ki} - p_{ki}, \quad i = 1,\ldots,L,\ k = 1,\ldots,M, \tag{8}$$

$$s_k = \sum_{i=0}^{L} s_{ki}, \qquad k = 1,\ldots,M, \tag{9}$$

$$\sum_{i=0}^{L} \theta_{ki} = 1, \qquad k = 1,\ldots,M, \tag{10}$$

$$p_i, p_{ki}, s_{ki} \geq 0,\ s_{k0} = 0,\ \theta_{ki} \in \{0,1\}. \tag{11}$$

Constraint set (2) enforces the assumption that each customer maximizes surplus, s_k. This is done by requiring that the customer surplus in the optimal solution be greater than or equal to the customer surplus that could be achieved through the purchase of any other bundle at the optimal prices. Constraints (3) are the subadditivity constraints. In (3) we write

$$B(i) = \bigcup_{j \in I} B(i), \tag{12}$$

instead of:

$$B(i) \subseteq \bigcup_{j \in I} B(i). \tag{13}$$

This is valid since it can be shown that the constraints in (13) with strict inclusions are redundant. Using strict set equality greatly reduces the number of constraints required in (3).

Because we prohibit explicit price discrimination between customer segments, if customer segment k buys bundle i and customer segment h buys the same bundle, then $p_{ki} = p_{hi}$. This condition is enforced by constraints (4) and (5). When $\theta_{ki} = 0$, then $p_{ki} = 0$ satisfies (4) and (5) since p_i need never exceed the maximum reservation price (over all customer groups and bundles). When $\theta_{ki} = 1$, then $p_{ki} = p_i$ and every customer segment buying bundle i pays price p_i. Constraints (6) are redundant to the underlying mixed integer programming problem but not to the linear programming relaxation. Because at most one θ_{ji} is positive (in the integer programming solution) for customer *j*, constraint (6) says that the surplus customer k receives for bundle i must be greater than or equal to the surplus possible from a bundle purchased by somebody else. This tightens the linear relaxation of the problem because the sum on the right-hand side of (6) is over *all* bundles, whereas the right-hand side in (2) includes only one bundle. Constraint (6) cannot replace constraint set (2) since it is enforcing customer surplus maximization for only those bundles actually purchased and does not include bundles which could be offered.

As every customer will purchase exactly one bundle, or will not make a purchase, this is a disjunctive constraint. The single purchase (or disjunctive) requirement is enforced by constraints (7)-(11), which involve the use of auxiliary variables. The key to understanding constraints (7)-(11) lies in constraint (10). This says that for each customer segment k exactly one θ_{ki} will equal one and the others zero. This implies that at most one p_{ki} and s_{ki} are ever nonzero. Thus constraints (7)-(9) correctly define the revenue and consumer surplus that firm and customer k get when bundle i is purchased.

If the objective function of (1) were optimized over the set of feasible points defined by (7)-(11) only, absolutely no enumeration would be required. Of course, other constraints such as (2) and (3) are present. Nonetheless, this modeling of the problem still has a tremendous effect on reducing the enumeration required, and thus greatly increases the size of problems which can be practically solved. For large number of components, Hanson and Martin have developed a relaxed bundle pricing model which only grows linearly with the number of components (not exponentially).

5 Summary

In this chapter, we have presented the traditional economic approach of profitable bundling and bundle pricing. Bundling ‚mechanisms' are perfectly pictorial in the single-firm/two-product case. Still, if the number of candidate bundles is low, the task to find the profit-maximizing strategy to assemble and price bundles can be carried out by hand. Nevertheless, trctable algorithmic solution procedures are generally needed, since the complexity of the optimization problem increases exponentially with the number of items to be bundled. Simulation techniques are gaining more and more importance as increasing computing power becomes available. It is true on the other side that optimization tools are deployed. The model of Hanson and Martin provided a programming framework which for the first time allowed to determine the optimal solution of the bundling problem analytically. Empirical testing and validation with real data however revealed significant influence of factors which are not sufficiently considered in the Hanson/Martin optimization model. These are:

- Continuous distribution of reservation prices: for product categories with a broad customer basis of heterogeneous preferences, segmentation according to the same reservation price structure may result in as many segments as customers exist. Collecting a small amount of ‚similar' segments through cluster analysis and assigning a stochastic reservation price variable (ideally following a ‚simple' distribution pattern such as the normal distribution) to them appears much more suitable.

- Bundle correlation: many product line items which are to be bundled appropriately have a complementary or substitutive influence on each other which must be accounted for. If a computer dealer offers an attractive hardware package, soaring bundle sales are likely to cannibalize other computer models on his shelf which are offered individually or within less attractive bundles.
- External competition: competitors certainly affect the bundling moves of a company in a non-monopolistic business environment. In particular, competitive pricing can be a crucial factor deciding upon success or failure of a product or bundle.
- Fixed costs: the decision to include a product or a product package in a product line is general associated with a fixed one-time cost, due to engineering, testing, production preparation and various planning or structural activities. High fixed costs may limit the scope of a product line and hence favor pure bundling strategies.

In the subsequent chapter on ‚Stochastic bundling‘, we establish an optimization model which takes up all above mentioned issues.

References

Adams, W.J. and J. L. Yellen (1976). “Commodity Bundling and the Burden of Monopoly.” Quarterly Journal of Economics, Vol. 90, 475-498.

Allenby, G. and P. Rossi (1991). “Quality Perceptions and Asymmetric Switching Between Brands.” Marketing Science, Vol. 10, 185-204.

Bauer, H., A. Herrmann and A. Mengen (1994). “Eine Methode zur gewinnmaximalen Produktgestaltung auf der Basis des Conjoint Measurement.” Zeitschrift für Betriebswirtschaft, Vol. 64, 81-94.

Blackstone, E. A. (1975). “Restrictive Practices in the Marketing of Electrofax Copying Machines and Supplies: The SCM Corporation Case.” Journal of Industrial Economics, Vol. 23, 189-202.

Burstein, M. L. (1960a). “The Economics of Tie-In Sales.” Review of Economics and Statistics, Vol. 42, 68-73.

Burstein, M. L. (1960b). “The Theory of Full-Line Forcing.” Northwestern University Law Review, Vol. 55, 62-95.

Carbajo, J., D. De Meza and D. J. Seidman (1990). “A Strategic Motivation for Commodity Bundling.” Journal of Industrial Economics, Vol. 38, 283-298.

Cattin, P. and D. R. Wittink (1989). “Commercial Use of Conjoint Analysis: An Update.” Journal of Marketing, Vol. 53, 91-96.

Cready, W. M. (1991). “Premium Bundling.” Economic Inquiry, Vol. 29, 173-179.

Daganzo, C. (1997). Multinomial Probit. Academic Press, New York.

Dansby, R. E. and C. Conrad (1984). “Commodity Bundling.” American Economic Review, Vol. 74, 377-381.

Dobson, G. and S. Kalish (1988). “Positioning and Pricing a Product Line.” Marketing Science, Vol. 7, 107-125.

Dobson, G. and S. Kalish (1993). “Heuristics for Pricing and Positioning a Product-Line Using Conjoint and Cost Data.”Management Science mm, Vol. 39, 160-175.

Eppen, G.D., W. A. Hanson and R. K. Martin (1991). “Bundling - New Products, New Markets, Low Risk.” Sloan Management Review, Vol. 32, 7-14.

Gaeth, G. J., I. P. Levin, G. Chakraborty and A. M. Levin (1991). “Consumer Evaluation of Multi-Product Bundles: An Information Integration Analysis.” Marketing Letters, Vol. 2, 47-57.

Garcia, C. B. and W. I. Zangwill (1981) Pathways to Solutions, Fixed Points and Equilibria. Prentice-Hall, Englewood Cliffs (N.J.).

Gossen, H. H. (1854). Entwicklung der Gesetze des Menschlichen Verkehrs und der daraus fliessenden Regeln für Menschliches Handeln. F. Vieweg, Braunschweig.

Green, P. E. (1984). "Hybrid Models for Conjoint Analysis: An Expository Review."Journal of Marketing Research, Vol. 21, 155-169.

Green, P. E. and A. M. Krieger (1992). "An Application of a Product Positioning Model to Pharmaceutical Products." Marketing Science, Vol. 11, 117-132.

Guiltinan, J. P. (1987). "The Price Bundling of Services: A Normative Framework." Journal of Marketing, Vol. 51, 74-85.

Hanson, W. A. and R. K. Martin (1990). "Optimal Bundle Pricing." Management Science, Vol. 36, 155-174.

Hanson, W. A. and R. K. Martin (1994). "Optimizing Multinomial Logit Profit Functions." Graduate School of Business, University of Chicago.

Jackson, B. B. (1985). Winning and Keeping Industrial Customers. Lexington (MA).

Kalish, S. and P. Nelson (1988). An Empirical Evaluation of Multiattribute Utility and Reservation Price Measurement. Purdue University.

Kamakura, W. and G. Russell (1989). "A Probablistic Choice Model for Market Segmentation and Elasticity Structure." Journal of Marketing Research, Vol. 26, 379-390.

Kohli, R. and R. Krishnamurti (1987). "A Heuristic Approach to Product Design." Management Science, Vol. 33, 1123-33.

Kohli, R. and R. Sukumar (1990). "Heuristics for Product-Line Design Using Conjoint Analysis." Management Science, Vol. 36, 1464-1478.

Lewbel, A. (1985). "Bundling of Substitutes." International Journal of Industrial Organization, Vol. 3, 101-107.

Louviere, J. and G. Woodworth (1983). "Design and Analysis of Simulated Consumer Choice or Allocation Experiments: An Approach Based on Aggregate Data." Journal of Marketing Research, Vol. 20, 340-367.

Manski, C. (1975). "Maximum Score Estimation of the Stochastic Utility Model of Choice." Journal of Econometrics, Vol. 3, 205-228.

McAfee, R. P., J. McMillan and M. D. Whinston (1989). "Multiproduct Monopoly, Commodity Bundling, and Correlation of Values." The Quarterly Journal of Economics, Vol. 104, 371-383.

McBride, R. D. and F. S. Zufryden (1988). "An Integer Programming Approach to the Optimal Product Line Selection Problem." Marketing Science, Vol. 7, 126-140.

McFadden, D. (1974). "Conditional Logit Analysis of Quantal Choice Behavior." Frontiers in Econometrics, (P. Zarembka ed.), Academic Press, New York.

McFadden, D. (1980). "Econometric Models for Probabilistic Choice Among Products." Journal of Business, Vol. 53, 13-34.

McFadden, D. (1986). "The Choice Theory Approach to Market Research." Marketing Science, Vol. 5, 275-297.

Nagle, M. (1987). The Strategy and Tactics of Pricing. Prentice-Hall, Englewood Cliffs (NJ).

Ogawa, K. (1987). "An Approach to Simultaneous Estimation and Segmentation in Conjoint Analysis." Marketing Science, Vol. 6, 66-81.

Palfrey, T. R. (1983). "Bundling Decisions by a Multiproduct Monopolist with Incomplete Information." Econometrica, Vol. 51, 463-483.

Porter, M. E. (1985). "Competitive Advantage: Creating and Sustaining Superior Performance." The Free Press, New York, 425-436.

Salinger, M. (1995). "A Graphical Analysis of Bundling." Journal of Business, Vol. 68, 85-98.

Schmalensee, R. (1982). "Commodity Bundling by Single-Product Monopolies." Journal of Law and Economics, Vol. 25, 67-71.

Schmalensee, R. (1984). “Gaussian Demand and Commodity Bundling.” Journal of Business, Vol. 57, 211-230.

Simon, H. (1992). Preismanagement. Gabler Verlag, Wiesbaden, 442-458.

Sobel, R. (1981). IBM-Colossus in Transition. New York, 1981.

Stigler, G. J. (1963). “United States vs. Loew's Inc.: A Note on Block Booking.” The Supreme Court Review, Vol. 152, 152-157.

Stigler, G. J. (1968). “A Note On Block Booking.” The Organization of Industry, (G.J. Stigler ed.), Irwin, Homewood (IL).

Tacke, G. (1989). Nichtlineare Preisbildung: Theorie, Messung und Anwendung. Gabler Verlag, Wiesbaden.

Telser, L. G. (1965). “Abuse in Trade Practices: An Economic Analysis.” Law and Contemporary Problems, Vol. 30, 488-505.

Warhit, E. (1980). “The Economics of Tie-in Sales.” Atlantic Economic Journal, Vol. 8, 81-88.

Wilson, L. O., A. M. Weiss and G. John. (1990). “Unbundling of Industrial Systems.” Journal of Marketing Research, Vol. 27, 123-138.

Yadav, M. S. and K. B. Monroe. (1993). “How Buyers Perceive Savings in a Bundle Price: An Examination of a Bundle's Transaction Value.” Journal of Marketing Research, Vol. 30, 350-358.

Zufryden, F. S. (1977). “A Conjoint-Measurement-Based Approach for Optimal New Product Design and Product Positioning.” Analytical Approaches to Product and Market Planning, (A.D.Shocker ed.), Marketing Science Institute, Cambridge (MA), 100-114.

Zufryden, F. S. (1982). “Product Line Optimization by Integer Programming.” Proc. Annual Meetings of ORSA/TIMS, San Diego (CA).

Stochastic Option Bundling and Bundle Pricing

Ralph Fuerderer[1], Arnd Huchzermeier[2], and Linus Schrage[3]

[1] **Ralph Fuerderer**, Adam Opel AG, International Technical Development Center, Ruesselsheim, Germany.

[2] **Arnd Huchzermeier**, WHU, Otto-Beisheim-Graduate School of Management, Vallendar, Germany.

[3] **Linus Schrage**, The University of Chicago, Graduate School of Business, Chicago, Illinois.

1 Introduction

In many production and services industries, *bundling* is the widespread practice of offering a number of products or services in a single package at an attractive price. In the automobile industry, a basic car model is offered along with options such as air conditioning, sun-roof, metallic exterior colors, and so forth, so-called free-choice or free-flow options. In general, the customer is able to purchase option packages which consist of a number of single options offered at a reduced bundle price. As far as most customer segments are concerned, equipment sales are the manufacturer's main source of profit. Thus, it is vital to decide on the pricing and the initial selection of free-choice options to be offered by the manufacturer. Due to substantial product and process development lead times, this task has to be carried out at least several months before production actually starts. Currently, accurate forecasting of demand for particular car types or option combinations is extremely difficult. The car manufacturer can hedge his risk of not matching the individual preferences of the customers for the car bundles offered, by providing a wide selection of free-choice options. However, from a manufacturing perspective, this product strategy is rather questionable. Moreover, variant-dependent costs are primarily determined by the number and the design of option combinations a customer can purchase with his basic car. Economies of scope exist among complementary options, e.g., a front door can be equipped with a power mirror more easily if it also has a power window. However, these cost synergies can only be exploited if customers do select (by chance) certain option combinations.

In order to maximize profit, the optimal price of a bundle requires a tradeoff between sales volume, profit margins and scale economies in manufacturing. In addition, cannibalization effects among competing options and bundles are determined by the size and the price vector of the entire supply of options. As the choice of the customer proves to be more price driven, intraline competition increases and thus the manufacturer responds by increasing the number of offered options and bundles. Consequently, the manufacturer's profitability can be negatively affected.

In this chapter, we demonstrate how a single firm can optimize the design and the price of its product line under uncertainty in reservation prices and consumer choice behavior considering both volume-dependent and variant-dependent costs. Furthermore, we assess two different bundling strategies: firstly, *pure bundling*, which is defined as the practice in which options that are already included in a bundle are not offered separately, and secondly, *mixed bundling*, which is defined as the practice in which no such restrictions are assumed. Although this paper mainly focuses on the automobile industry, the presented methods can be applied to almost any other industry.

In the following section, we present a brief review of the relevant literature. In section 3, we propose a method for determining both the bundle composition and the bundle prices under uncertainty. In addition, we map market specific customer choice patterns by employing a very general share-of-utility rule. On the cost side, we take variant-dependent costs into account. The overall problem is formulated as a mixed-integer non-linear program. In section 4, we show how to generate the necessary model input data very efficiently and comment on the role of the underlying cost structure. A solution algorithm based on a decomposition method is presented in section 5. For that purpose, we apply a path following technique to bypass poor local optima which exist due to the inherent non-concavity of the problem. In section 6, we test our procedure with a real large-scale data set of an European manufacturer in the automobile industry. Furthermore, we comment on implications for managers based on model experiments and an extensive sensitivity analysis. The paper concludes with a summary in section 7.

2 Literature Review

The existing literature reflects the impact of bundling with respect to law and politics, competition, and the demand side. Only few authors have carried out a formal economic analysis. In this paper, we comment briefly on the most influential studies and focus on the quantitative aspects, i.e., model assumptions and results.

In an earlier examination, Stigler (1963) reports on an antitrust case of a movie distributor, who is trying to raise profits by leasing movie packages which contain movies of varying attractiveness. In his analysis, Stigler introduces the concept of a reservation price. This concept reflects the demand of a discrete customer segment and represents the maximum price that consumers of this segment are willing to pay for a movie. Reservation prices are assumed to be additive, costs are not considered explicitly. A buyer's utility for a movie is represented by his surplus which is the difference of his personal reservation price and the real product price. If his surplus is positive, he will purchase the movie. Stigler concludes that this form of pure bundling, as a subtle tool of price discrimination, yields profit if reservation prices correlate negatively, but show similar aggregations (bundle reservation price) for each segment. Many subsequent studies have been carried out within the Stigler framework.

Adams and Yellen (1976) compare both pure and mixed bundling strategies with perfect price discrimination which is known to be the most profitable form of pricing. They examine a two-product monopolist and several consumer segments with additive reservation prices. Variable costs are assumed to be also additive and no fixed costs are considered. Their findings are based on the assumption that a customer will purchase exactly one unit of the surplus maximizing good. If all surpluses are negative, he will not purchase at all. Adams and Yellen point out that the two major factors which determine the profitability of either sales strategy are the level of cost and the distribution of customers in the reservation price space. For instance, a negative correlation of reservation prices favors bundling strategies as opposed to single pricing. With regard to a symmetric price distribution and a relatively high cost level, a single pricing strategy should be preferred.

For the first time ever, Schmalensee (1984) developed numerical criteria to decide on the profitability of a bundling strategy. In a two-product scenario, he assumes that the reservation prices of the buyers follow a bivariate normal distribution in which the additivity assumption for reservation prices is retained. He achieves conditions of profitability in either bundling case by using numerical analysis. Schmalensee also proves that a negative correlation of reservation prices is not a necessary condition for the profitability of pure bundling. Moreover, he generally suggests that mixed bundling is superior to pure bundling and unbundled sales.

The first authors providing a practical method for a single firm to design and to price optimal bundles are Hanson and Martin (1990). Hanson and Martin assume a monopolist facing segmented customer demand with subadditive reservation prices. Furthermore, volume-dependent costs are assumed to be subadditive, too. Variant-dependent costs are not considered at all. Again, the customer will choose the bundle which maximizes his surplus. The authors define the bundle pricing problem as a mixed-integer program in which the number of constraints increases exponentially with the number of single items. For reasons of tractability, they suggest a relaxed formulation with only linear growth in the number of constraints. Their solution method is based on a decomposition method and tested with a variety of generated data sets.

However, none of these papers show how to obtain optimal bundles given a long forecast horizon and uncertainty in reservation prices. The choice behavior of customers when confronted with several products and bundles generally follows a maximum utility pattern which does not allow for a choice of several products or product bundles within a product line. Variant-dependent costs reflecting scale economies are not taken into account either.

3 Model Formulation

In this section, we present a mathematical formulation of the stochastic bundle pricing problem. Firstly, we provide an overview of all model parameters and variables:

Demand Parameters:

r = the number of options, $k = 1,\dots,r$.

m = the number of candidate bundles, $j = 1,\dots,m$.

m_0 = maximum number of bundles permitted.

n = the number of customer segments, $i = 1,\dots,n$.

N_i = number of customers in segment *i*.

α_i = choice parameter for segment *i*, $\alpha = (\alpha_1,...,\alpha_n)$.

l_j = competitive lower price bound of bundle *j*, $l = (l_1,...,l_m)$.

u_j = competitive upper price bound of bundle *j*, $u = (u_1,...,u_m)$.

μ_{ij} = mean reservation price of segment *i* for bundle *j*.

σ_{ij} = standard deviation of reservation price of segment *i* for bundle *j*.

φ_{ij} = reservation price density function of segment *i* for bundle *j*.

φ_{ij} = inverse reservation price distribution function of segment *i* for bundle *j*.

Production Parameters:

c_{ij} = variable costs of supplying one customer of segment *i* with bundle *j*.

f_j = fixed costs associated with bundle *j*.

$\rho_{ijj'}$ = linear correlation of bundle *j* with bundle *j'* for segment *i*, $\rho_{ijj'} \in [0;1]$.

$$a_{kj} = \begin{cases} 1 & \textit{if option } k \textit{ is contained in bundle } j \\ 0 & \textit{otherwise} \end{cases}$$

$A_j = \{a_{1j},...,a_{rj}\}$ (bundle j); $\mathbf{A} = \text{Mat}(a_{kj})$.

Decision Variables:

p_j = the price of bundle *j*.

$$y_j = \begin{cases} 1 & \textit{if bundle } j \textit{ is produced} \\ 0 & \textit{otherwise} \end{cases}$$

Auxiliary variables:

$\mathbf{p} = (p_1,...,p_m)$.

$\mathbf{y} = (y_1,...,y_m)$.

ε_{ij} = consumer surplus of segment *i* for bundle *j*.

π_{ij} = purchase probability of segment *i* for bundle *j*.

Π_i = profit/customer obtained from segment *i*.

Π = profit function $N_1 \Pi_1 + ... + N_n \Pi_n$.

The model is based on the following assumptions:

Model Assumptions:

1. Reservation prices for segment i for bundle j is assumed normally distributed.
2. Demand correlation of bundles j and j' for segment i isindependent of **p** and **y**. Due to technical reasons, the correlations are mapped onto the unit interval by the linear transformation $t \to \frac{1-t}{2}$.

3. The utility of bundle j for segment i only depends on the realized consumer surplus

$$\varepsilon_{ij}(p_j) = \int_{p_j}^{\infty} (p - p_j)\varphi_{ij}(p)dp \tag{1}$$

and the supply of other substitutive bundles.

4. Intraline competition for segment i and bundle j can be approximated by a modified α-choice pattern, i.e.,

$$\pi_{ij}(\mathbf{A}, \mathbf{p}, \mathbf{y}, \alpha_i) = \frac{\varepsilon_{ij}^{\alpha_i} y_j}{\sum_{j'=1}^{m} \rho_{ijj'} \varepsilon_{ij}^{\alpha_i} y_{j'}} \tag{2}$$

5. No customer undertakes a purchase resulting in a negative consumer surplus for each bundle. Furthermore, customers do not adopt a „mental accounting scheme“ where they maximize the surplus for a predefined budget.

6. All customer segments face the same bundle prices.

In practice, a normal distribution seems to be a good fit for the distribution of observed customer preferences of options and bundles. Furthermore, a normal distribution does not preclude negativereservation prices expressing strong dislikes for a bundle over manyother distributions, e.g., the lognormal or exponential distribution which is restricted to positive reservation prices only. Numerical procedures can be used for efficient evaluation of cumulative distributions and loss functions. However, the first assumption which has been stated above does not provide a necessary condition for the feasibility of the presented method.

In assumption 2, the linear correlations between two bundles do not depend on the decision variables. Empirical examinations of historical car data confirm that despite of substantial price changes linear correlations among options vary only within small margins over time.

Similar to the deterministic Hanson-Martin model, consumer surplus is defined as the normal linear loss function. However, this does not provide a full explanation of a customer segment's utility for a bundle. Hence, maximum consumer surplus does not automatically imply a customer's decision to purchase a bundle. Choice behavior is also influenced by the price and design of other bundles. In assumption 4, segment-specific cannibalization effects among competing bundles are represented by a modified α-choice rule (Green and Krieger, 1992). Roughly speaking, to obtain the choice probability of bundle *j* for segment *i*, the surplus $\varepsilon_{ij}(p_j)$ from purchasing bundle *j* competes with all offered substitutive bundles of the product line. The degree of competition is expressed by the linear correlation $\rho_{ijj'}$. Perfectly substitutive bundles ($\rho_{ijj'} = 1$) enter into (2) in terms of their full surplus. Perfectly complementary bundles ($\rho_{ijj'} = 0$) do not compete with bundle *j*. In addition, each surplus term is equipped with a segment specific exponent α_i measuring the inherent degree of randomness in consumer choice behavior. Given that all bundles are produced (y = (1,...,1)) and are perfectly substitutive ($\rho_{ijj'} = 1$), we obtain for

$$\pi_{ij}(\mathbf{A}, \mathbf{p}, \mathbf{y}, \alpha_i = \infty) = \begin{cases} 1 & \text{if } \varepsilon_{ij}(p_j) = \max\{\varepsilon_{i1}(p_1), \ldots, \varepsilon_{im}(p_m)\} \\ 0 & \text{otherwise} \end{cases} \tag{3}$$

which corresponds to the maximum utility choice rule deployed in the Hanson-Martin model. If $\alpha_i = 1$, (2} is equivalent to the linearly weighted Terry-Bradley-Luce choice rule, and for $\alpha_i = 0$, we obtain the totally random choice behavior with $\pi_{ij}(\mathbf{A},\mathbf{p},\mathbf{y},0) = 1/m$. Thus, the model is able to adapt to the market-specific product choice behavior by allowing for the adjustment of the value of α_i. In Figure 1, we illustrate the behavior of $\pi_{ij}(\mathbf{A},\mathbf{p},\mathbf{y},\alpha_i)$ for n = 1, m = 4, y = (1,...,1), perfectly substitutive bundles, and surpluses $(\varepsilon_{11},\ldots, \varepsilon_{14}) = (5,6,7,9)$.

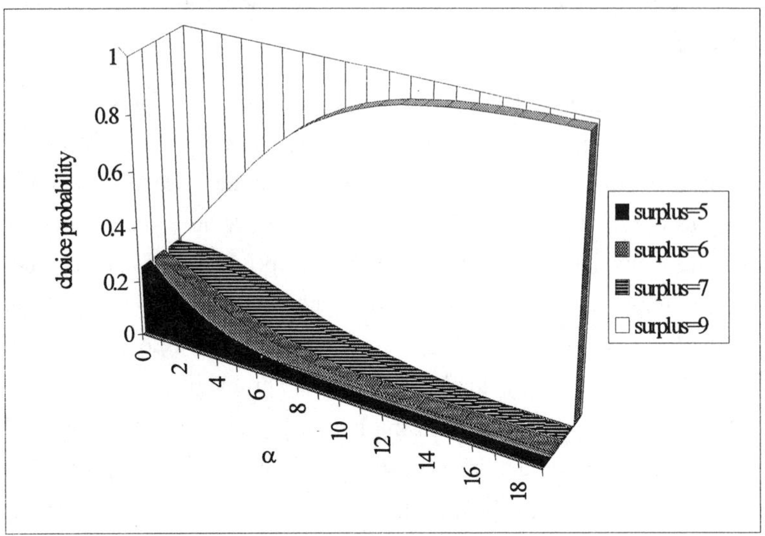

Fig. 1. Alpha choice rule

Given that all offered bundles show perfect complementarity, the share of buyers in segment i for bundle j equals $\varphi_{ij}(p_j)$. Under any other linear correlation matrix among bundles, $\varphi_{ij}(p_j)$ provides an upper bound for the penetration of bundle j. Assumption 5 states that a customer will refuse to purchase a bundle if it does not provide him with a positive utility. In assumption 6, explicit price discrimination among different customer segments is prohibited.

Without losing generality, it is assumed that the first r bundles are identical to the respective r single options. We then formulate the stochastic bundling problem in (4) to (9), given assumptions 1 through 6.

Objective

$$\max -\sum_{j=1}^{m} f_j y_j + \sum_{i=1}^{n}\sum_{j=1}^{m} N_i (p_j - c_{ij}) \pi_{ij}(\mathbf{A}, \mathbf{p}, \mathbf{y}, \alpha_i) \varphi_{ij}(p_j) \,, \tag{4}$$

subject to

$$\sum_{j=r+1}^{m} y_j \le m_0 \,, \tag{5}$$

$$l_j \le p_j \le u_j \qquad \text{j} = 1, ..., \text{m} \,, \tag{6}$$

$$y_j \in \{0,1\} \qquad j = 1,\ldots,m, \tag{7}$$

plus for the Pure Bundling strategy

$$a_{jk}(y_k + y_i) \leq 1 \qquad k = 1,\ldots,r;\, j = r+1,\ldots,m, \tag{8}$$

or plus for the Mixed Bundling strategy

$$\max_{k \leq r, r+1 \leq j \leq m} a_{kj}(y_k + y_j) \geq 1. \tag{9}$$

In the profit function (4), variable and fixed costs for each bundle and segment are considered. The share of bundle j at price p_j for segment i is computed as π_{ij} $(\mathbf{A},\mathbf{p},\mathbf{y},\alpha_i)$ φ ij (pj).

Due to the non-linear share function, the objective is a mixed-integer non-linear function. Constraint (5) enforces that the maximum number m_0 of multi-option bundles ($j > r$) is not exceeded. For reasons of competitive pricing, optimal prices are bounded by constraint (6). The upper bound u_j captures the intention to underprice an equivalent bundle of a different manufacturer. A lower bound l_j may be necessary to meet profitability requirements set by specifically financial plans. In case of a pure bundling strategy, no single option is offered which is already contained in an optimal bundle, see constraint (8). In a mixed bundling strategy, constraint (9) enforces that at least one option is available both as single option and within a bundle.

Although production restrictions on option combinations may apply, the number of potential bundles m grows exponentially with the number r of options Considering all feasible bundles, an exponential number of constraints is required. In section 4, possibilities of restricting the candidate bundles to a moderate size prior to the optimization are discussed. Furthermore, we comment on the role of the deployed cost structure. Also, an efficient procedure of how to obtain estimates for the reservation price distribution, the linear correlation matrix of each segment and the choice parameter α is suggested.

4 Input Data Generation

The stochastic bundling problem requires several types of input data.

4.1 Candidate Bundles and Bundle Correlation

A large number of bundles may neither yield cost synergies nor enhance demand. In order to obtain a tractable mathematical program (see section 3), it is reasonable to limit the computations to a pool of candidate bundles from which the set of optimal bundles is chosen. A common method to exclude unprofitable bundles a priori is an expert survey. Besides experimentally based methods, as conjoint analysis, analytic methods to obtain part worth estimates may be of particular interest. Linear correlations among candidate bundles can be easily obtained by using historical sales information on option basis and applying standard linear regression methods. The resulting linear bundle correlation matrices $\mathbf{R}_i = (\rho_{ijj'})$ are quadratic, unsymmetric and exhibiting ones on the main diagonal.

4.2 Customer Segments, Price Distributions and Choice Behavior

For reasons of efficiency, our model requires aggregation of individual consumer preferences. Thus, cluster analysis can be used in which the tradeoff between loss of information and computational tractability from large segment sizes should be considered with caution. In the automobile industry, for instance, it is convenient to cluster customers according to their choice of engine size, trim level, or body style. Historical data can be used to develop clustering criteria which lead to similar equipment purchases within the segments. This procedure is used to obtain a moderate number of segments and to limit the loss of information through aggregation.

In order to obtain an estimate for the moments of the normal price distributions and values for α_i, we propose a least squares approach using historical sales data. It needs to be emphasized that historical sales data does not necessarily reflect customer preferences. The purchase decision of a customer is frequently based on compromises and is also influenced by the effectiveness of the sales force. In the US, for example, customers prefer to buy new cars from the dealer's showroom. In Europe, large fractions of newly built vehicles are made to order. In our choice model, we collect sales data for each segment i and bundle j at times t, t=1,...,T. Let p_{jt} denote the bundle price observed at time t, and s_{ijt} the respective option share. Hence, we run a regression to obtain estimates for μ_{ij}, σ_{ij}, and α_i utilizing

$$\sum_{j=1}^{m}\sum_{t=1}^{T}\left(\frac{\varepsilon_{ij}^{\alpha_i}\left(p_{jk}\right)}{\sum_{j=1'}^{m}\rho_{ijj'}\varepsilon_{ij}^{\alpha_i}\left(p_{j'k}\right)}\varphi_{ij'}\left(p_{j'k}\right)-s_{ijt}\right)^2 \tag{10}$$

where the functions ε_{ij} and φ_{ij} are expressed in terms of μ_{ij} and σ_{ij} as previously defined.

Again, expert opinions need to be employed if either new options are assessed (no historical data exists) or truncated price data is not sufficient to obtain a sharp minimum (little price variation over time). Figure 2 shows a characteristic error parabola for the proposed least squares method if we fix the choice parameters α_i.

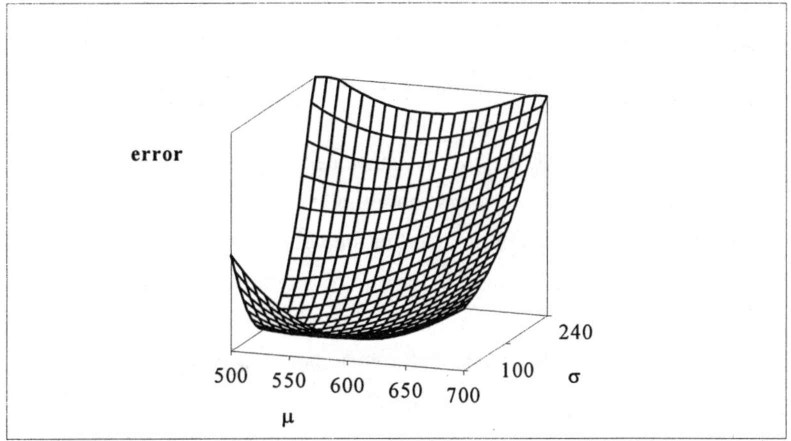

Fig. 2. Error distribution of least square method

Alternatively, we could also minimize the sum of positive relative errors

$$\left|\frac{\varepsilon_{ij}^{\alpha_i}\left(p_{jk}\right)}{\sum_{j'=1}^{m}\rho_{ijj'}\varepsilon_{ij}^{\alpha_i}\left(p_{j'k}\right)}\varphi_{ij'}\left(p_{jk}\right)-s_{ijt}\right|s_{ijt}^{-1} \tag{11}$$

However, the least squares approach has shown better convergence rates.

4.3 Costs

The model cost structure includes variable costs which may vary across customer segments as well as variant-dependent fixed costs. Variable costs c_{ij} include costs for material and labour on one side. On the other side, c_{ij} consist of variable variant-dependent costs which are assumed linear. In the example taken from the automobile industry, variable variant-dependent costs result from changing work contents at one work station due to assembly variants, increased transportation effort and extended storage space due to feature variety, among others. In distribution, e.g., dealers are planning their stock according to product variety and expected sales. Variant-dependent fixed costs f_j are typically encountered in the fields of engineering and testing, production, marketing and sales. If, e.g., a front seat containing heating, position memory, height adjustment and side-airbags has to be developed, engineering and testing costs do not depend on production volumes. By the same token, tooling and investment costs as well as costs for advertisement can be considered non-variable. Explicit treatment of these cost schemes is beyond the scope of this chapter. However, most car producers are able to estimate and assign variant-dependent complexity costs from their activity based cost methods. Moreover, the cost structure depends on the bundling strategy employed. In a mixed bundling strategy, complexity costs for a bundle tend to be low because the single options are included in the vehicle program independent of the bundling strategy. Complexity costs for a pure bundle are significantly higher because the aggregated costs for the options included in the bundle have to be imposed. Thus, it is desirable to adapt the cost data according to the bundling strategy pursued. Cost data collection is the hardest task of input data generation. This is particularly relevant if complex development, manufacturing and distribution processes are involved.

5 Solution Method

Classic non-linear programming models for product line selection and pricing problems are proven to be NP hard (see Dobson and Kalish, 1988). Thus, we decompose the stochastic bundling problem in an integer master problem and a non-linear subproblem.

Master Problem (M):

Objective

$$\max \; -\sum_{j=1}^{m} f_j y_j + \theta \qquad (12)$$

subject to integer constraints (5) and (7) – (9)

Subproblem (S):

$$\theta = \max \sum_{j=1}^{m}\sum_{i=1}^{n} N_i\left(p_j - c_{ij}\right)\pi_{ij}\;(\mathbf{A},\mathbf{p},\mathbf{y},\alpha_i)\,\varphi_{ij}\left(p_i\right) \qquad (13)$$

subject to linear constraint (6).

In order to solve the master problem (M), a greedy interchange heuristic (see Cornuejols, Nemhauser and Fisher, 1978) is used. At each solution step of (M), a new bundle from the bundle candidate pool is added to the set of optimal bundles and the subproblem (S) is executed. The heuristic solution method is described as follows:

1. Add a feasible bundle A_{j*} from the candidate bundles to the optimal set of bundles $A_{j1},\ldots, A_{jm'}$.
2. Solve (S) for $m'+1$ bundles $A_{j1},\ldots,A_{jm'}, A_{j*}$.
3. Memorize objective of (M) and repeat step 1, until all candidate bundles are checked.
4. Add profit maximizing bundle to optimal bundles $A_{j1},\ldots,A_{jm'}$ and increase m by 1.
5. If $m' < m_0$ goto step 1, else terminate.

For a fixed value of m', step 2 is carried out for all remaining candidate bundles. In the worst case, this step is carried out $m - m' + 1$ times. However, solving the non-linear subproblem (S) at each step proves extremely difficult, since

$$\Pi\;(\mathbf{A},\mathbf{p},\mathbf{y},\alpha) = \sum_{i=1}^{n}\sum_{J=1}^{m} N_i\left(p_j - c_{ij}\right)\pi_{ij}\;(\mathbf{A},\mathbf{p},\mathbf{y},\alpha_i)\,\varphi_{ij}\left(p_j\right) \qquad (14)$$

may be non-noncave, depending on the empirically observed values for the choice behavior α. Consequently, the necessary condition for the existence of a global extreme point (maximum) – which is expressed by the concept of concavity – may not be fulfilled. The following simple example illustrates that (15) is not even quasi-concave, in general. Consider a single consumer with $\alpha = 3.0$, and three competing bundles ($y = (1, 1, 1)$, $\rho_{ijj'} = 1$) with $(\mu_1, \mu_2, \mu_3) = (500, 100, 500)$, and $(\sigma_1, \sigma_2, \sigma_3) = (150, 30, 200)$. Set marginal volume-dependent costs $c_{lj} = 0$ for all bundles, and assume price vectors $p^l = (600, 50, 500)$ and $p^2 = (500, 80, 600)$. Resulting profits are $\Pi(\mathbf{A}, \mathbf{p}^1, \mathbf{y}, 3.0) = 238.98$ and $\Pi(\mathbf{A}, \mathbf{p}^2, \mathbf{y}, 3.0) = 226.56$, but profits from a combined price vector are $\Pi(\mathbf{A}, 1/2\,(\mathbf{p}^1 + \mathbf{p}^2), \mathbf{y}, 3.0) = 188.91$. Even for small values of α, the following Lemma shows that we can expect concavity of (15) only on a subset of the feasible price region.

LEMMA 1: *Assume without loss of generality that* $\mathbf{y} = (1,...,1)$. *Then, there exists* $\varepsilon > 0$, *such that for all* α *with* $\alpha_i \leq \varepsilon$ *holds that* $\Pi(A, \mathbf{p}, y, \alpha)$ *is concave, as long as*

$$l_j \leq p_j \leq \min\left(u_i, \frac{\mu_{ij} - c_{ij}}{2} + \sqrt{\left(\frac{\mu_{ij} - c_{ij}}{2}\right)^2 + 2\sigma_{ij}} \right) \tag{15}$$

for all $j \leq m$.

Proof: See Fuerderer (1996).

Hence, if a bundle price p_j exceeds $\mu_{ij} + \sqrt{2\sigma_{ij}}$, concavity can not be guaranteed and conventional solution methods may get stuck in poor local maxima. In order to avoid this pitfall, we employ a path following technique which was in a similar way presented by Hanson and Martin (1994). The key idea is to define a smooth transition or *homotopy* between the original non-concave objective function and a concave approximation function. For an introduction to homotopy theory, see Aleksandrov (1965), Schubert (1969) or Garcia and Zangwill (1981). Starting with the well-known global maximum of the approximation, we drag our path along the maxima of the intermediate functions with a small constant step size (see Figure 3).

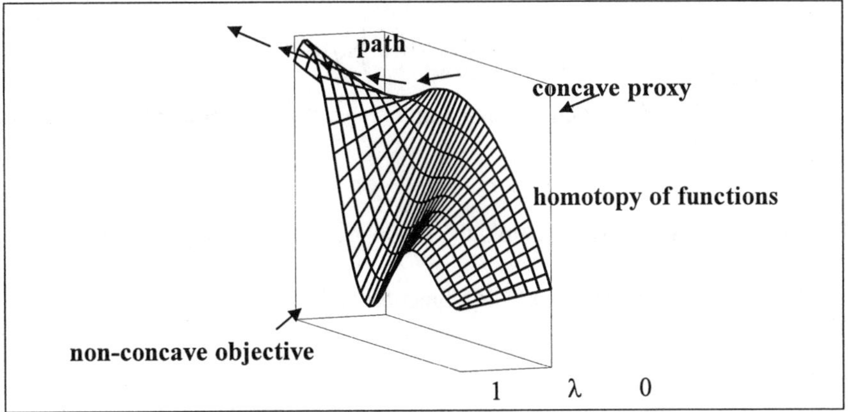

Fig. 3. Path following method

Similar to the idea of Hanson and Martin (1994), we are aiming at a well-behaved approximation of the alpha choice function π_{ij} (**A**, **p**, **y**, α_i). Braibant and Fleury (1985) suggest a convex linearization which can easily be reversed to a concave linearization.

In order to improve computational tractability, we will choose a low degree polynomial function as an approximation instead. In the modified choice function, demand for bundle j is decoupled from demand for other bundles. To maintain a proper scaling, the modified choice function and the original coincide at the upper price bound **u**. For a fixed reference price vector $\mathbf{r} = (r_1,\dots,r_m)$, we define

$$\bar{\Pi}_{ij}(p) = \frac{1}{\varphi_{ij}(p_j)(p_j - c_{ij})}\left(a_2 p_j^2 + a_1 p_j + a_0\right), \tag{16}$$

where

$$\begin{aligned} &a_2 < 0. \\ &a_1 = -2a_2 r_j. \\ &a_0 = (u_j - c_{ij})\pi_{ij}(u_j) - a_2 u_j^2 - a_1 u_j. \end{aligned} \tag{17}$$

The magnitude of a_2 is not yet determined and can be used to scale the approximation appropriately. The fixed parameters **A**, **y**, and α are here omitted for ease of notation. We now define a homotopy as a linear combination of the original profit function and the profit function with modified choice behavior as follows.

$$H(\lambda.p)=\sum_{i=1}^{n}\sum_{j=1}^{m}N_i\left(p_j-c_{ij}\right)\left(\lambda\Pi_{ij}+(1-\lambda)\bar{\Pi}_{ij}\right)\varphi_{ij}\left(p_j\right) \tag{18}$$

for $\lambda \in [0;1]$. The following Lemma summarizes the properties of the concave approximation.

LEMMA 2: With the above notation, the following properties hold:

1. $H(\lambda, \mathbf{p})$ is continuous on $[0;1]$ as a function of λ
2. $H(1, \mathbf{p}) = \Pi(\mathbf{A}, \mathbf{p}, \mathbf{y}, \alpha)$,
3. $H(1, \mathbf{u}) = H(0, \mathbf{u})$,
4. $H(0, \mathbf{p})$ *it is strictly concave and attains its global maximum at* $\mathbf{p} = \mathbf{r}$.

Proof: See Fuerderer (1996).

Our next goal is to show that the extremal behavior of any member $H(\lambda, \mathbf{p})$ of the homotopy class is similar to $H(\lambda + \varepsilon, \mathbf{p})$ if $\varepsilon > 0$ is small enough. A necessary condition for the existence of an extreme point is that all partial derivatives with respect to price are zero. Our goal is to locally characterize the behavior of extremal points as functions of λ, as we increase λ to $\lambda + \varepsilon$ for all $\lambda \in [0;1)$. Unfortunately, the system

$$\nabla_p H(\lambda, p) = 0 \tag{19}$$

is non-linear for all $\lambda \in (0; 1]$. Thus, it is not possible to solve the system directly. However, we can describe the local behavior of (20) near any regular point.

LEMMA 3: *Let $\lambda_0 \in I=(0, 1)$, $\mathbf{p}_0 = (p_1^0,\dots,p_m^0)$ a price vector, such that $\nabla_p H(\lambda_0, \mathbf{p}^0) = 0$. Furthermore, let $Hess(H(\lambda, \mathbf{p}))$ be regular at (λ_0, p^0). Then there is a $\varepsilon > 0$, continuously differentiable functions h_j on $(\lambda_0 - \varepsilon; \lambda_0 + \varepsilon) \cap I$, such that $p_j^0 = h_j(\lambda_0)$ and $\nabla_p H(\lambda, h_1(\lambda),\dots,h_m(\lambda)) = 0$ for all $\lambda \in (\lambda_0 - \varepsilon, \lambda_0 + \varepsilon) \cap I$.*

Proof: Lemma 3 follows immediately from the implicit function theorem, for instance, see Dieudonné (1969).

Lemma 3 implies that if we drag from one function of the homotopy to another for a sufficiently small step size, the extreme points do not change significantly as long as the Hessian matrices are regular. However, the lemma is not constructive in the sense that it does not render an explicit representation of the local resolutions h_j. Also, it does not show whether we can choose a ε independent of λ as a global step size. In the next corollary, we demonstrate that, provided we avoid singular points, there will always be a constant step size $\bar{\varepsilon} > 0$ such that our algorithm terminates after $\left[\frac{1}{\bar{\varepsilon}}\right]+1$ steps.

COROLLARY 4: *Let $(\lambda_k, \mathbf{p}^k)$ be a sequence of extreme points of H with $\lambda_k < \lambda_{k+1}$ for all positive integers k. Then there exists $\lambda_0 < 1$ with $\lambda_k \rightarrow \lambda_0$, $p_0^j = lim_{k\rightarrow\infty} p_j^k$ exists for all $j \leq m$ and Hess $(H(\lambda, \mathbf{p}))$ at $(\lambda_0, \mathbf{p}^0)$ is singular for $\mathbf{p}^0 = (p_1^0,...,p_m^0)$.*
Proof: see Fuerderer (1996).

This demonstrates that no strictly decreasing sequence of step sizes can exist without violating the regularity condition mentioned above. Another potential pitfall are bifurcation points which can cause our path search to be misdirected to a poor local optimum of *H(1)*.

Let S = $\{(\lambda, \mathbf{p}) \mid \nabla_p H(\lambda, \mathbf{p}) = 0\}$. $(\lambda_0, \mathbf{p}^0)$ is called a *bifurcation point of the equation* $\nabla_p H(\lambda, \mathbf{p}) = 0$, if (1) $(\lambda_0, \mathbf{p}^0) \in S$ and (2) for all neighborhoods U of $\mathbf{p}^0$, Λ of λ_0 and $\mathbf{p}^1, \mathbf{p}^2 \in U$, we have $\mathbf{p}^1 \neq \mathbf{p}^2$ and $(\lambda_0, \mathbf{p}^1), (\lambda_0, \mathbf{p}^2) \in S$. If we run into a bifurcation point during a search step it is impossible to decide which path leads to the global maximum. The following bifurcation theorem tells us that again bifurcation points can only be singular points of $H(\lambda, \mathbf{p})$.

THEOREM 6: *Let $\Lambda = (\lambda_k - \varepsilon, \lambda_k + \varepsilon) \cap (0,1)$ for a $\lambda_k \in (0,1)$ and $\varepsilon > 0$. Let U be a neighborhood of $\mathbf{p}^k$ and $\nabla_p H(\lambda_k, \mathbf{p}^k) = 0$. Let the Hessian of $H(\lambda_k, \mathbf{p}^k)$ be continuous at $(\lambda_0, \mathbf{p}^0) \in \Lambda \times U$ and let $\mathbf{p}^0$ be a regular point of H. Then, $(\lambda_0, \mathbf{p}^0)$ is no bifurcation point.*
Proof: see Fuerderer (1996).

Thus, any point $(\lambda_0, \mathbf{p}^0)$ can only be a bifurcation point of $\nabla_p H(\lambda, \mathbf{p}) = 0$ if 0 is an eigenvalue of Hess $(H(\lambda, \mathbf{p}))$ at $(\lambda_0, \mathbf{p}_0)$. We know now that once we have found a maximum of one of our profit functions $H(\lambda)$, the corresponding maximal price vector in most cases provides a good starting point for the optimization of a subsequent profit function $H(\lambda + \varepsilon)$. Since we are not protected against encountering singular points, it has to be stressed that there is no guarantee for our optimization approach finding the global maximum of our initial profit function.

A possibility to strengthen the presented method is to vary the step size and the not yet specified reference price vector **r**. If several choices of (ε, **r**) lead to the same resulting optimal price vector for (15), we can be quiet sure that we are truly optimal.

6 Computational Results

We have implemented the proposed heuristic algorithm on several computer systems. At each step of the path following procedure, we used a gradient projection method based on Rosen's Method (see Rosen (1960), (1961)). All employed derivatives are computed analytically. Line searches are performed utilizing the golden section method.

The test data reflects projected sales of a mid-size car model of the Adam Opel AG for the German market. The car is available as a 3-door hatchback, a 4-door notch/5-door hatchback, and a 5-door wagon. The different body styles turned out to be an appropriate clustering of the customers with regard to their habits of option purchase.

We have selected ten comfort/safety options, e.g., anti-lock brakes, electric sun-roof, power door locks, etc., and designed twelve bundles, composed of at most five options. These bundles both reflect economies of scope and potentially high market share. In the next step, we determined the reservation price distribution for each of the three segments and all single options and bundles. Best fits for the choice parameters are 0.686, 0.689 and 0.832, respectively, which confirms a relatively high degree of randomness in option choice behavior. With this initial data the choice model approximates historical option sales penetrations according to (11) with a mean relative error of 5.4%.

Finally, we determined the variable bundle costs as well as variant-dependent costs utilizing variant tree techniques and activity based costing. In order to estimate the fixed costs associated with a bundle, we need to distinguish between a mixed and a pure bundling strategy. Since each option can still be ordered separately, there are no stock reduction or deproliferation savings for mixed bundling. Thus, the fixed costs for a bundle were estimated as the sum of fixed costs of the included single options. For pure bundling, we expressed the pooling and process deproliferation effect in terms of bundle j fixed costs f_j by

$$f_j = \left(\sum_{k=1}^{r} a_{kj} f_k^2 \right)^{\frac{1}{2}} \qquad j = 1, \ldots, m. \tag{20}$$

The optimization procedure was run on a 486 Siemens-Nixdorf PC for a mixed and a pure bundling strategy taking into account up to $m_0 = 5$ common optimal bundles for all segments. To evaluate the quality of the results, we used total enumeration instead of the greedy heuristic and implemented it on a CRAY YMP 2-32. Figure 4 shows all relevant problem parameters and effort/quality measures.

Both bundling strategies show a certain trend for hierarchy. An optimal bundle for a small number of permissible bundles m_0 often proves also optimal for a strategy with larger values for m_0. This characteristic is more pronounced for pure bundling than for mixed bundling because in a mixed bundling strategy each bundle not only competes with other bundles but also with all single options that are part of the bundle. This characteristic supports the utilized greedy heuristic which tends to preserve each bundle which was recognized as optimal in prior runs. Cornuejols, Fisher and Nemhauser (1978) obtained a worst case quality gap of approximately 37% between greedy solutions and the true optimum. In our calculations, the quality gap was always less than 8%. This finding strengthens the appropriateness of a greedy open heuristic in this context.

Number of Pure Bundles	CPU time (secs.)	Quality of Optimum in %
1	4,1	100,0
2	16,8	100,0
3	64,5	99,1
4	169,4	92,1
5	298,7	94,2
Number of Mixed Bundles	**CPU time (secs.)**	**Quality of Optimum in %**
1	8,8	100,0
2	38,8	100,0
3	101,9	100,0
4	222,1	98,0
5	386,2	94,3

Fig. 4. Problem parameters for the stochastic bundle optimization

In Figure 5 we show the maximum profit obtained by each bundling strategy as a function of the number of bundles.

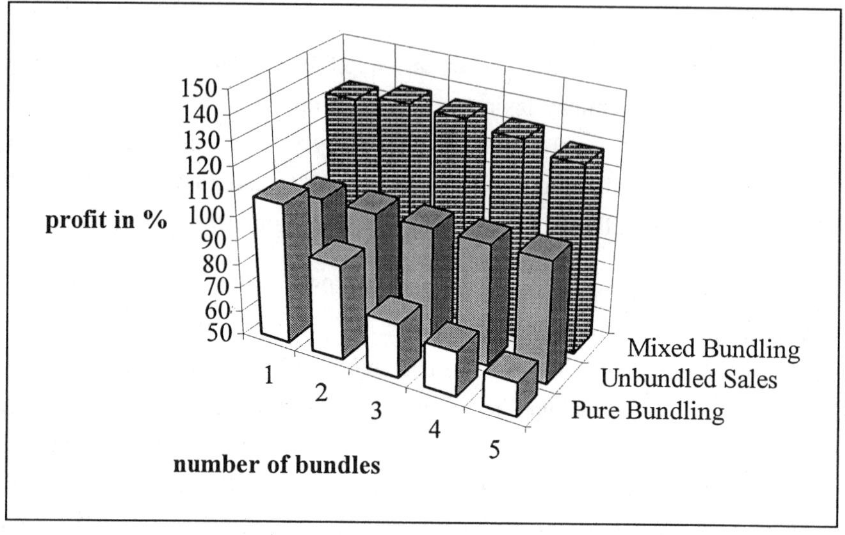

Fig. 5. Maximum profit for different numbers of bundles

The most important finding is that although it does not reduce proliferation costs at all, mixed bundling can be considered as the most profitable bundling strategy. The profit obtained by two optimal bundles is approximately 34% higher than for unbundled sales. Moreover, profitability of pure bundling turned out to be more sensitive with regard to the maximum number of bundles m_0 than profits obtained by mixed bundling. This happens because all candidate bundles contain hot seller options which yielded high single option profit before. Penetrations for the pure bundles are, however, lower in such a way that the lost single option profit can not be outweighed by higher demand for low runner options within a bundle. Nevertheless, one optimal pure bundle is still 5% more profitable than unbundled sales.

For both strategies, the resulting optimal prices are usually decreasing as the number of optimal bundles increases. The proposed discounts for the bundles vary between 15 and 35% of the single option prices. Although optimal prices for comparable bundles are slightly higher in a pure bundling strategy, penetrations of bundles are higher due to the missing competition with free-choice options.

6.1 Sensitivity Analysis

We have performed a Monte-Carlo simulation to assess the dependency of the potential profits from each bundling strategy on the randomness of choice behavior. For this reason, we randomly generated 500 α-values for each of the three segments that were equally distributed in the interval [0.5; 1.5]. Demand structures of various car segments which have been analyzed could be described by α values falling in the selected range. As an aggregated measure for the choice randomness, we took the Euclidean norm of the three αs:

$$\bar{\alpha} = \sqrt{\alpha_1^2 + \alpha_2^2 + \alpha_3^2} \tag{21}$$

Hence, $\bar{\alpha}$ is approximately normally distributed with a mean of $\sqrt{3}$ and a standard deviation of 0.44. It proves rather interesting that maximum profit for mixed bundling is increasing as choice behavior gets less random, while, at the same time, maximum profit for pure bundling is decreasing. This is shown in Figure 6. In general, the resulting changes are small and almost linear, while the optimal number of bundles for each strategy is not affected at all.

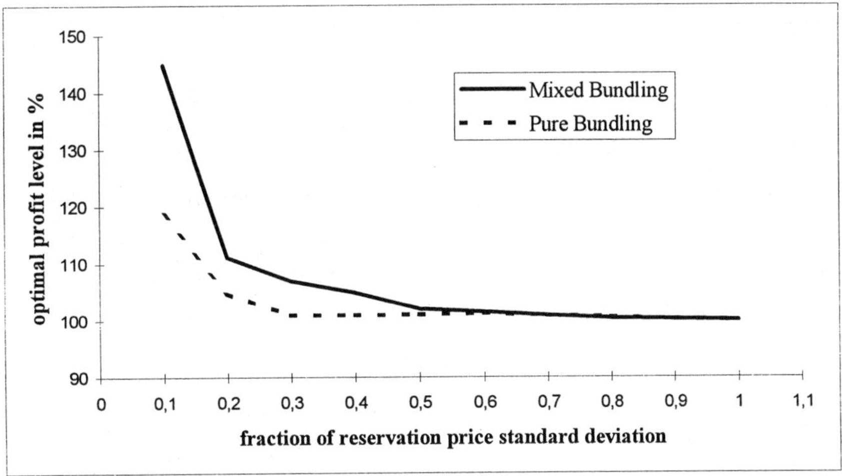

Fig. 6. Sensitivity of profit with respect to choice randomness

It is also interesting to find out how optimal profit changes with decreasing forecast horizon or increasing similarity in purchasing behavior within a segment. For this reason we have repeated our computations by sequentially reducing reservation price standard deviations by 10%. In Figure 7, we observe that reduced forecast horizon increases optimal profit for pure and mixed bundling with a more significant effect on mixed bundling.

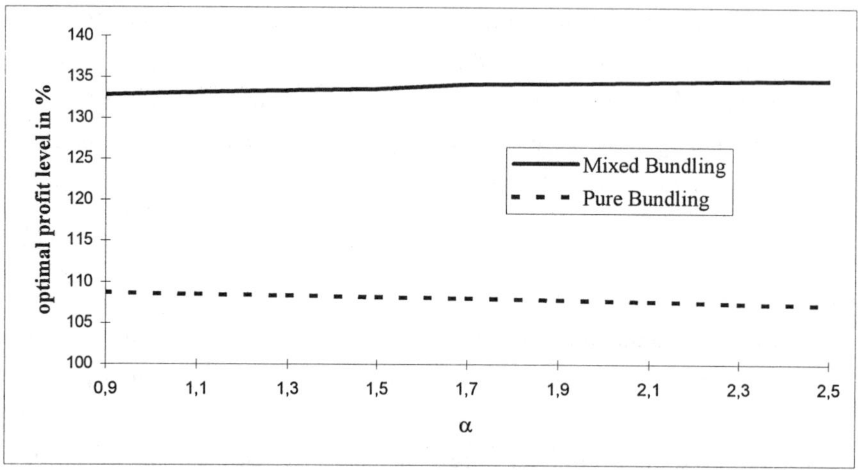

Fig. 7. Sensitivity of profit with respect to increased forecast horizon

However, uncertainty of reservation prices has to be reduced significantly, e.g., an impact on optimal profit can not be observed before "0.6σ" for mixed bundling, respectively "0.3σ" for pure bundling.

6.2 Case Study Results

Eppen, Hanson and Martin (1991) provide a list of management rules applying to the practice of bundling. Our simulations with real data overall confirm their insights, but moreover reveal a couple of new opportunities. Although analytical optimization is inevitable to fully exploit a pricing strategy, the following empirical findings may serve as managerial rules-of-thumb in order to select the best pricing scheme.

<u>Single Pricing:</u>

A single pricing strategy proves effective in a situation where a product line of limited scope has to be priced. In particular, if reservation prices exhibit a strong positive correlation and considerable variance, a single pricing strategy enables to extract surpluses separately for each product by adapting prices for the single items. Hence, confirming results of Adams and Yellen (1976), Schmalensee

(1984) and Salinger (1995), a bundling strategy can hardly overcome the well-known extraction problems in the positively correlated case. Similarly, in the case of rather high cost levels with little opportunity for economies of scope, single pricing is in many cases superior to bundled sales. As bundle prices can not be substantially decreased versus the single item prices, no additional demand can be created, and customers purchasing the bundle are pulled over from corresponding single item sales (cannibalization) at overall lower margins.

Bundle Pricing:
In most situations when one of the above discussed single pricing requirements is not met, bundled sales provide potential for lowered costs and enhanced demand. In case of rather high profit margins as well as in a range of negative to moderate positive price correlation, bundling is in general more profitable than a single pricing strategy. By adjusting the bundle discount, additional demand through improved surplus extraction can be optimally traded off with shrinking profit margins. Regarding scale economies, bundle pricing provides an opportunity to increase the sales volume of one product by tying it to successful products within a package (this argument also applies to the situation when a new product with uncertain market performance is introduced). In addition, cost synergies among complementary products (economies of scope) can balance out the effect of sub-additive profit margins for the bundle.

However, complexity cost reductions can only be fully exploited under a pure bundling strategy, which aims at collecting heterogeneous demand with a minimum dispersion. In the framework of global economies of scope and substantial deproliferation, pure bundling can be a clearly more powerful pricing and planning tool than mixed bundling, extending all earlier published results.

Mixed bundling turns out to be an immensely valuable strategy if long forecast horizons have to be dealt with, as demonstrated in section 6.1. As already noted also, if products exhibit large standard deviations, mixed bundling is the optimal strategy to target the bundle to an aggregate market, but to extract additional surplus through higher priced individual items from unusual customer segments. Moreover, we find mixed bundling strictly more profitable than pure bundling in the following situations. First, if cannibalization effects prevail, comprehensive bundles must be "backed up" by an array of single items to avoid severely dropping demand shares. Second, in strongly unsymmetric cases (e.g., standard deviations of products differ greatly) mixed bundling enables to differentiate prices of a bundle and individually offered items of large price uncertainty.

General Remarks:
As the high predictability of our model is traded off against substantial problem complexity, our above guidelines clearly have to follow a *ceteris paribus* pattern. In a real-world situation, a detailed assessment on the appropriateness of either pricing strategy and its implementation should only be conducted through a formal analytical optimization as presented in this paper.

7 Summary and Model Extensions

In this paper, we have surveyed and analyzed various bundling strategies focusing on the automobile industry. For the first time, a model has been developed which enables manufacturing managers and design engineers to find profit maximizing pure bundling and mixed bundling policies under uncertainty. The model has been formulated as a mixed-integer non-linear program which also takes into account fixed costs. We have presented a decomposition method to solve the stochastic bundling problem and applied a path following technique to tackle non-concave properties. Poor local maxima have to be dealt with if traditional non-linear methods are used. We have tested the methodology with topical and complex datasets from an European car manufacturer and assessed the impact of changes of key input data, e.g., forecast horizon, on profit. Furthermore, we translated our experiences gained from numerical analyses into management guidelines.

Although our methodology is not restricted to certain industrial branches, it has other limitations. We have presented the situation of a single firm which can only react to prices set by competition. We have, however, ignored the global impact of a bundling strategy under competition. It requires game theoretic modelling approaches to incorporate brand switching criteria, for instance. It is also plausible to use column generation techniques to improve the efficiency of the greedy interchange algorithm. Even if we need to limit the number of considered candidate bundles, analytic selection methods before and during the optimization can enhance the computational effort even more.

More insights about the stochastic aspects of product bundling have to be gained. Since the complexity of our models limit the possibilities to answer all questions by solving the problems directly, we have to rely on numerical experiments.

References

Adams, W. J. and J. L. Yellen (1976). "Commodity Bundling and the Burden of Monopoly." Quarterly Journal of Economics, Vol. 90, 475-498.

Aleksandrov, P. (1965). Combinatorial Topology. Graylock Press, Baltimore (MD).

Bazaraa, M.S., H. D. Sherali and C. M. Shetty (1993). Nonlinear Programming. Wiley Interscience Series in Discrete Mathematics and Optimization, New York.

Braibant, V. and C. Fleury (1985). "An Approximation Concept Approach to Shape Optimal Design, Computer Mehtods." Applied Nechanics and Engineering, Vol. 53, 119-148.

Cornuejols, G., M. L. Fisher and G. L. Nemhauser (1977). "Location of Bank Accounts to Optimize Float: An Analytic Study of Exact and Approximate Algorithms." Management Science, Vol. 23, 789-810.

Dieudonné, J. (1969). Foundations of Modern Analysis. Academic Press, New York.

Dobson, G. and S. Kalish (1993). "Heuristics for Pricing and Positioning a Product-line Using Conjoint and Cost Data." Management Science, Vol. 39, 160-175.

Eppen, G.D., W. A. Hanson and R. K. Martin (1991). "Bundling – New Products, New Markets, Low Risk." Sloan Management Review, Vol. 32, 7-14

Fuerderer, R. (1996). Optimal Component and Option Bundling under Demand Ridk – Mass Customization Strategies in the European Automobile Industry. Gabler, Wiesbaden.

Garcia, C. B. and W. I. Zangwill (1981). Pathways to Solutions, Fixed Points and Equilibria. Prentice-Hall, Englewood Cliffs (N.J.)

Green, P. E. and A. M. Krieger (1992). "An Application of a Product Positioning Model to Pharmaceutical Products." Marketing Science, Vol. 11, 117-132.

Hanson, W. A. and R. K. Martin (1990). "Optimal Bundle Pricing." Management Science, Vol. 36, 155-174.

Hanson, W. A. and R. K. Martin (1994). "Optimizing Multinomial Logit Profit Functions." Graduate School of Business, University of Chicago.

Kohli, R. and R. Sukumar (1990). "Heuristics for Product-Line Design Using Conjoint Analysis." Management Science, Vol. 36, 1464-1478.

Rosen, J. B. (1960). "The Gradient Projection Method for Nonlinear Programming, Part I, Linear Constraints." SIAM Journal of Applied Mathmatics, Vol. 8, 181-217.

Rosen, J. B. (1961). "The Gradient Projection Method for Nonlinear Programming, Part II, Nonlinear Constraints." SIAM Journal of Applied Mathmatics, Vol. 8, 514-532.

Schmalensee, R. (1984). "Gaussian Demand and Commodity Bundling." Journal of Business, Vol. 57, 211-230.

Schubert, H. (1969). Topologie. Teubner Verlag, Stuttgart.

Stigler, G. J. (1963). "United States vs. Loew's Inc.: A Note on Block Booking." The Supreme Court Review, Vol. 152, 152-157.

Bundling and Pricing of Modular Machine Tools Under Demand Uncertainty

Nils Tönshoff[1], Charles H. Fine[2], and Arnd Huchzermeier[3]

[1] **Nils Tönshoff**, BMW AG, Muenchen, Germany.

[2] **Charles H. Fine**, MIT, Sloan School of Management, Cambridge, Massachusetts.

[3] **Arnd Huchzermeier**, WHU Koblenz, Otto-Beisheim Graduate School of Management, Vallendar, Germany.

1 Introduction

In the early nineties, the global machine tool industry was struck by an economic downturn in most of its customers' industries, especially the automotive industry and its suppliers. Consequently, a loss of more than 50% of production resulted in the German machine tool industry, see Figure 1 and the statistical reports published by the VDMA (1993, 1994b, 1995). German manufacturers were severely hit and many suffered losses that drove them out of business or forced them to merge with other manufacturers. Well known in the German industry are the Deckel merger with MAHO and afterwards the merger of Deckel-MAHO with Gildemeister in mid 1994. Overall, global competition between machine tool manufacturers has become fiercer. Recently, the German Machinery and Plant Manufacturers' Association warned that it was almost impossible to generate profit by building machine tools in Germany, see VDMA (1991).

German manufacturers were loosing market share measured in terms of machine exports, see VDMA (1993) since 1974 before gaining share again in 1984. In 1986, the U.S. market share was smaller than the German share and ever since a similar difference of 2-5% between the shares remained. The Japanese market share has been rising with decelerating speed and has been at the level of the U.S. share since 1986, see Figure 2 (Source: VDMA (1993)). From 1991 on, the shares decreased because Portugal, Spain and Singapore, South Korea, Taiwan and Hong-Kong were included in the "major exporting countries."

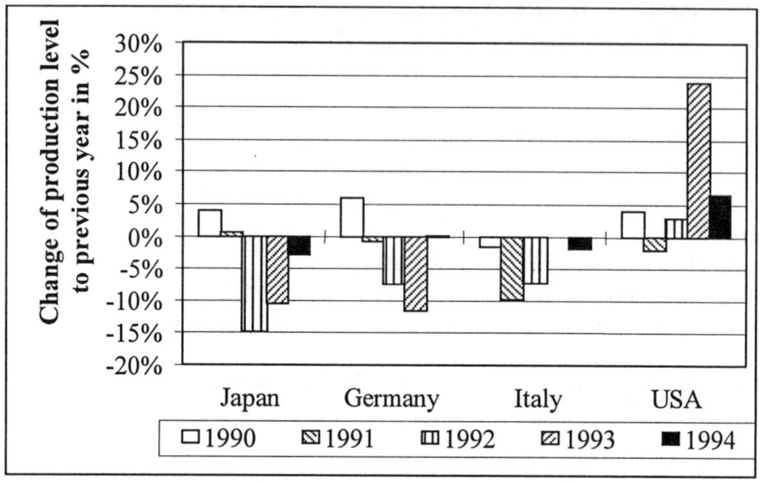

Fig. 1. Change of production in major machine tool producing countries

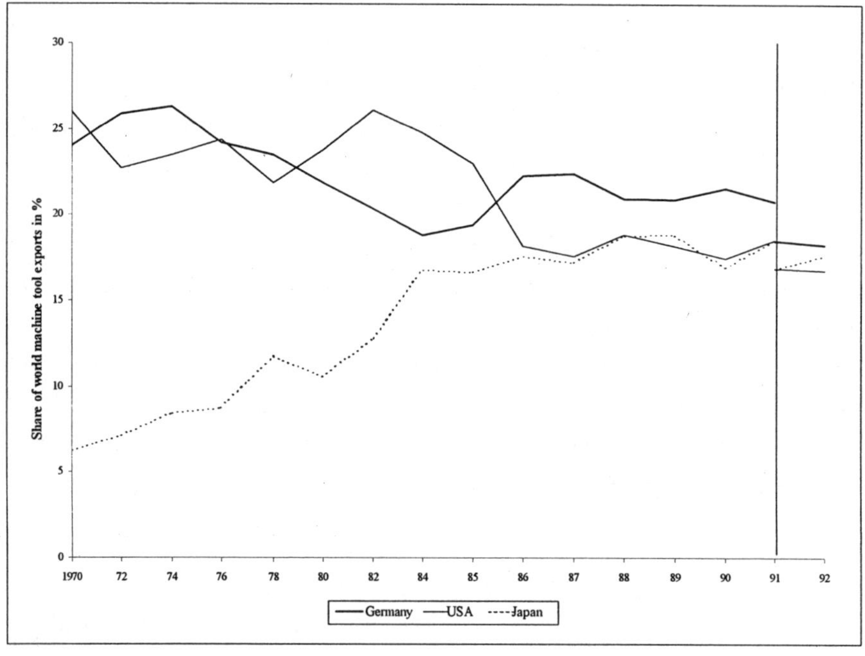

Fig. 2. Share of global exports of machines

Increasingly, machine tool users demand customized machine tools that fit their specific needs, due to shorter product life cycles and higher levels of customization. Increased competition in the machine tool industry leads to highly price-conscious customers requesting multi-functionality of products. Two main product strategies have emerged from this market situation. Japanese volume producers exploit the markets for standard machinery successfully. They have implemented a cost/price leadership strategy driven largely by scale economies. In response, German and Swiss manufacturers followed a strategy of product differentiation and concentrated on the market for high-priced and customized machine tools.

However, more Japanese manufacturers are seeking inroads to the high-end markets. In order to maintain their relative market share, German machine tool manufacturers are challenged to accelerate in product and process innovations to lower costs even further. In the German machine tool industry, we have observed a major conflict between production and marketing which is caused by the prevailing sequential planning and decision process. Marketing wants to offer customized products with high performance at low prices. Production tries to accommodate these demands through multi-functionality, but often delivers "over-engineered" products that exploit scale economies in manufacturing and assembly only. However, this does not correspond to the requirement by marketing to manufacture exactly to customers' specifications.

In order to overcome this dilemma, modular manufacturing has been adopted widely in the German machine tool industry, e.g., see Spur (1996). This allows manufacturers to offer a high degree of customization at relatively competitive prices. However, a modular product structure does not necessarily lead to profit. Moreover, through reverse engineering competitors can quickly erode any innovation advantage. Thus, short product life cycles and price competition prevail in the global machine tool industry.

Alternatively, a *product bundling* and *pricing* strategy allows the machine tool builder to exploit the willingness-to-pay of customers for particular product options and thus raise product prices. Also, the distribution of demand can be used to enhance the profitability of the firm and to lower delivery time or to enhance the service level significantly.

The remainder of this chapter is organized as follows. In section 2, an overview of the related literature is provided. The notion of a modular machine tool is developed (2.1) and the reasons and problems of overly complex product lines with high variety (2.2) are described. The dilemma of standardization on one side and customization on the other, is developed (2.3-2.5) and modularization is proposed as a strategy to overcome it (2.4). Section 3 criticizes the modularization strategy. In order to implement a bundling and pricing strategy, a number of decisions have to be made which are outlined in section 4. An integrated model formulation is proposed. Overall, the model has a non-linear structure. However, the mixed-integer non-linear model can be reformulated as a mixed-integer linear model accounting for discrete demand scenarios. Two application case studies of leading firms in the German machine tool industry are presented in section 5. Section 6 gives some insights to the implementation of the model using the case study data. Finally, section 7 contains concluding remarks.

2 Traditional Approaches to Variety Management

In the following, select readings in the engineering, marketing and operations literature pertaining to product variety management are discussed briefly.

2.1 Modeling Machine Tool Manufacture

In general, a machine tool can be viewed as the combination of physical parts that, when working together properly, provide the machine tool user with a number of desired functions. The process of modularization of machine tools means to combine different functions of machine tools to modules, e.g., see Tönshoff and Böger (1994). In what follows, we assume that the machine tool contains modules and that modules “contain” functions.

In the engineering literature, models of modular machinery have been studied extensively. Jovane (1989) models an abstract elementary machine that can be used as a model for machine tools. Building that model he takes into account the specific information on the product, machine, process and environment that needs to be contained in an elementary machine model. Tönshoff and Böger (1994) develop a reference architecture that is used to describe functions and modules of a machine tool. Their goal is to establish a machine model that allows to describe machines and their functionality independent of the physical embodiment of the units. They envision a higher degree of division of labor in which a company specializes in producing only modules of a machine tool. Having established standardized interfaces and joints between modules, the modules of different manufacturers can be combined to machine tools.

In general, the number of assembly operations decreases as the number of components that have to be mounted onto the machine stand is reduced. On the other hand, the integration of functions into a single component may require additional design work and may lead to higher manufacturing costs, e.g., see Ulrich et al.(1993). These phenomena are referred to as super- or subadditivity of costs. According to practitioners in machine tool design and design researchers, subadditivity is the most common case in practice. However, a trade-off between design costs and the costs of the "risk of design" needs to be considered. The latter stands for the increased risk and cost of a growing number of prototypes and of potential errors and failures in testing and producing the more complex and highly integrated modules.

The theoretical background of standardizing engineering products has received great attention by mechanical engineers, e.g., see Pahl and Beitz (1984, 1993), but has also been dealt with in the economics literature, e.g., see Pindyck and Rubinfeld (1992) and Gabel (1991). Tönshoff and Böger (1994), in a machine tool user study, find that the standardization of interfaces between subsystems of a manufacturing system is one of the six top requirements of future machine tools. Franke et al. (1995) show in a case study that the cost savings of standardizing machine tools can be attributed primarily to areas that generate fixed overhead costs, i.e., research and development, design, production planning, supply chain logistics and inventory management. Hamel (1996) describes the separation of form and function as one of the key strategic factors that successful companies employ.

2.2 Variety and Complexity

Since customers have been demanding high degrees of customization ever since, mainly German manufacturers have provided the marketplace with highly specialized but expensive machines. Overall, this leads to a product variety proliferation problem. Roever (1992), Quelch and Kenny (1994) and Hardle et al. (1994) study the effects of overly complex products and product lines. They come to the conclusion that especially for consumer goods the proliferation of products can lead to increased costs and lower profitability.

Kekre and Srinivasan (1990), on the other hand, show in an empirical study of more than 1,400 business units that a broader product line does not necessarily lead to decreased profitability or to increased manufacturing costs. This is due to manufacturing strategies adopted by the firms, e.g., the introduction of manufacturing cells and group technology, the reduction of set-up times that facilitates Just-in-Time delivery or flexible manufacturing. Other strategies include the integration of marketing, manufacturing and logistics and finally, there are product strategies that could have been adopted and that might explain the rather counterintuitive results of this study, e.g., the utilization of parts commonality or the adoption of a postponement strategy, see also Lee et al. (1993).

Stalk (1988) and Stalk and Hout (1990) also state that increased product variety and better service does not necessarily lead to lower profitability, but can provide the manufacturer with an opportunity of significant price premia in the global market. The authors describe means that allow for higher variety and service level while maintaining or even lowering costs. Some of the time-based strategies are also discussed in Blackburn (1991). Relating all the results and examples to the machine tool industry, one can state that due to the heterogeneous demand a machine tool manufacturer faces, a diverse product line is a necessity. Thus, he can only strive to mitigate the negative effects that might be incurred through this diversity, using for instance, some of the strategies mentioned above.

Since the acquisition of machine tools requires high investments, the customer is only willing to pay for the functions he needs. One reason for the steep decline and loss of market share of the German machine tool industry lies in the German machine tool containing too many functions. These cannot be utilized by most customers and cause the machine to be too expensive. This phenomenon has been characterized as "*German over-engineering*", see Jorissen et al. (1993). An explanation for machine tool manufacturers including too many functions in a machine is their response to the demand uncertainty for functions. This is caused partly by long innovation cycles and by frequent changes of the requirements of the machine tool user.

2.3 Standardization

As described before, the machine tool manufacturer faces a dilemma: on the one hand the customer wants a customized machine tool that satisfies all requirements and on the other hand he wants this customized product to be as low priced as a standardized machine tool. In order to lower costs and offer more attractive products at lower prices, product standardization has been proposed as a viable approach. A simple standardization strategy will not solve the problem since it only concentrates on reducing the number of different parts without considering implications for manufacturing and marketing aspects. McCutcheon et al. (1994) describe this dilemma as "The Customization-Responsiveness Squeeze". A machine tool manufacturer copes with the following demand environment: customers demand quick delivery of totally customized products. These products can be customized in final stages of production if the manufacturer adopts the appropriate manufacturing strategy. According to McCutcheon et al. (1994), this environment calls for an "assemble-to-order" approach. A certain inventory risk for the pre-produced parts exists, though. In machine tool manufacturing the costs of R&D as well as design costs represent a major part of the costs (7.4% of total costs in 1990, according to a report published by the VDMA (1991)). In 1992, the expenditures for R&D amounted to 3.8% of the sales volume, see VDMA (1994a). It is a well-known fact that during the design stage of a product, most of the costs of the final product are determined, e.g., see Pahl and Beitz (1993). In the machine tool industry, an adapted "assemble-to-order" approach is to pre-design the subsystems of a machine and to produce them to stock. The final product is then assembled-to-order.

2.4 Modularization

The business strategy that allows a manufacturer of machine tools to achieve low-cost customization to a certain degree is a modularization strategy, see Starr (1965), Pine (1993), Reiß and Beck (1994) and Ullrich and Eppinger (1995). In order to offer machine tools that yield some scale economies but that are still giving a customer almost exactly what he wants, machine tools are offered as modular products.

Ulrich and Tung (1991) describe modularity as a relative property depending on the similarity between the physical and functional architecture of the components and the number of incidental interactions between physical components.

In a recent study carried out by the Critical Technologies Institute, RAND, on the global machine tool industry, especially on the largest producers of machine tools in Japan, Germany and the U.S., one of the findings demonstrated that Japanese machine tool producers gained a competitive advantage through the creation of modular components, see Finegold et al. (1994a). This modularization of products leads to a productivity edge of Japanese machine tool manufacturers. By cutting the number of different parts and employing scale effects, CNC machines tools could be made affordable to any customer, see Finegold et al. (1994b). The study urges Western manufacturers to follow the Japanese example. There are examples of German and U. S. machine tool builders that have already implemented a modular product strategy, e.g., see Malle (1994) and the interview with Dr. A. Herrscher of INDEX-Werke GmbH & Co. KG, a major German machine tool manufacturer, in the Siemens-Zeitschrift (1995).

Potential benefits and costs of modularity have been described in a case study by Ulrich and Tung (1991). Benefits other than the already mentioned scale economies are the ease of a product change, ability to offer a great variety of products, order lead time reductions, the decoupling of tasks in design, production and testing, a design and production focus and the ease of product diagnosis, maintenance, repair and disposal. Costs of modularity can be caused by the still rather static product architecture of the modular product, by the redundancy found in modular designs and by the excess capability of components that can be used for different tasks. The greatest risk involved is that of being copied by competitors more easily. This is due to using modular components because functions and interconnections are very clear and well defined.

2.5 Mass Customization

The goal achieved by the revised "assemble-to-order" approach is the same as the goal of a "mass customization" strategy. Pine (1993) explains that by designing and building standardized subproducts, scale economies are exploited. These subproducts are combined in the final production step to quasi-customized products, thus exactly satisfying the customer's demand and therefore leading to higher market share. Westbrook and Williamson (1993) describe mass customization as carefully controlling complexity so that many of the mass production techniques are still applicable. This is why true individuality and high variety have to be balanced cautiously. Even though Womack (1993) criticizes some of the means and anticipated positive effects of mass customization, the fundamental idea of combining the positive manufacturing and marketing effects of standardization and customization is captivating. Reiß and Beck (1994) show ways of achieving the goal of mass customization for the volume producer as well as for the make-to-order manufacturer. Volume producers should focus more on customer requirements through modularization of products while keeping their cost advantages. The make-to-order manufacturer should maintain his customization advantage and focus on cost reductions through standardizing components. Kotha (1994) argues that neither mass production nor mass customization bring about optimal results but that the pursuit of both strategies provides the company with greater strategic flexibility. These results are exemplified in the case studies of the National Bicycle Industrial Co. by Kotha and Fried (1993) and Fisher (1994).

3 Bundling and Pricing Strategy

Overcoming the shortcomings of a mere modularization strategy, the *bundling strategy* aims at three different goals that are increasing both the value of the machines for the customers and the profit to the selling firm: i) cost reductions through standardization of modules and consideration of demand risk, thus avoiding expensive dis- and re-assembly, ii) shorter delivery lead times achieved through complexity reduction and iii) higher prices through the bundling of options, i.e., modules and through the skimming of consumer surplus.

A simple example of how *higher prices* can be achieved through the bundling of modules is the following. Consider a machine tool and two options that can be included in the machine, an *on-line programming system* and an *automatic part-handling system*. Customers are assumed to have reservation prices they are willing to pay for the options. If the option is offered for a price that is lower than the customer's reservation price for it, the customer is going to consider buying the option. Deciding which options to buy, the customer maximizes his consumer surplus. This is defined as the difference between market price and reservation price. Thus, if the price for a good is high but the customer's reservation price is low he tends not to buy the product. On the other hand, if the customer places a high value on the good, i.e., his reservation price is high, he will buy the product even if the market price is high, see Figure 3. The two axes span the customer reservation price space. Reservation prices for both options are assumed to be positive. In this case, the machine tool manufacturer divides the customer space into four quadrants by setting certain prices for the two options, e.g., the price for the programming system p_{PS} and the price for the automatic part-handling system p_{APH}. Customers who fall in quadrant *I* have reservation prices that are both higher than the prices for the options. Thus, these customers will buy both options. Customers in quadrant *II* have a reservation price for the programming system that is higher than the price demanded by the manufacturer but a reservation price for the automatic parts handling system that is lower than the price. This means that these customers will buy the programming system but not the automatic part handlig. Similarly, customers in quadrant *III* will buy the automatic part handling but not the programming system. Finally, customers in quadrant *IV* have reservation prices that are lower than the demanded prices. They will therefore not buy either option.

Bundling means to combine certain options into a bundle and sell this bundle for a bundle price p_B that normally includes a positive or negative discount. *Pure bundling* means to only offer options in a bundle. *Mixed bundling* means to bundle options but to also offer options not as part of a bundle. Pure bundling and setting a bundle price of p_B divides the customer space into a upper right half and a lower left half, e.g., see the dashed line in Figure 3. Customers in the upper right half have reservation prices that sum up to an amount that is higher than the bundle price. Thus, these customers will buy the bundle. Customers in the lower left half have a sum of reservation prices that is lower than the bundle price and as a consequence will not buy the bundle.

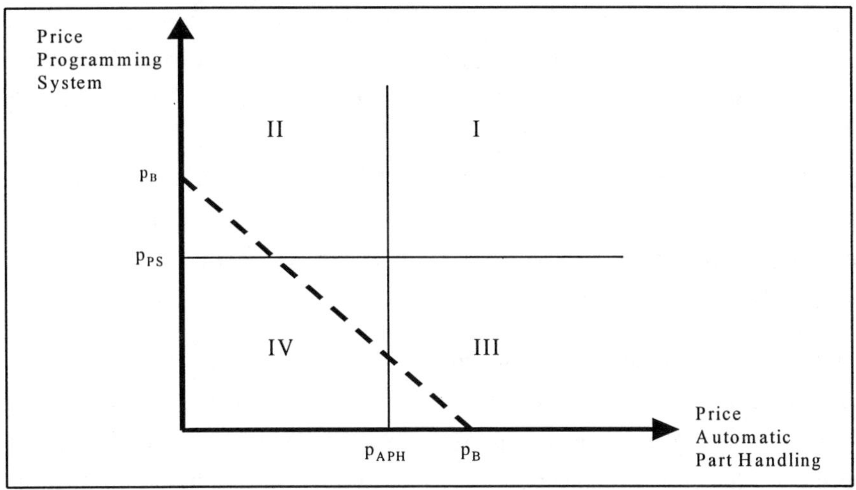

Fig. 3. Reservation prices and bundling

The notion of bundling was introduced by Stigler (1963) and has been dealt with in the economics and management science literature extensively. Salinger (1995) analyzes the effects of bundling on profitability and welfare. One of his results is that the consumer surplus can increase through bundling, if bundling leads to lower prices. Green and Krieger (1991) analyze how to design and bundle services in order to gain market share in certain aggregate markets or market segments. In the machine tool industry, customers can hardly be aggregated to segments since they have a very heterogeneous demand and are used to get customized products. Eppen et al. (1991) develop seven guidelines on how to optimally bundle products, most of which apply directly to the machine tool design problem, too. Reducing costs through increasing production efficiency, expanding markets and increasing demand through aggregating products and improving product performance for the customer through increased joint performance of bundles are some of the advantages of bundles described.

The optimal pricing of bundles has also been dealt with in the marketing literature, e.g., see Lilien, Kotler and Moorthy (1992). Green and Krieger (1985), McBride and Zufryden (1988), Kohli and Sukumar (1990) and Dobson and Kalish (1993) devise formulations and heuristics for the optimal product line selection employing conjoint analysis techniques. Lele (1992) explains that if customers do not have information about the components, the bundles can be priced like a new, unbundled product. In case customers have some or full knowledge, the bundle must have a price that is smaller than the sum of the prices of the single components. Alternatively, it must provide some additional quality like higher reliability or convenience. Thus, Lele concludes, bundling is valuable especially for new and innovative products that customers do not have much knowledge about. Pindyck and Rubinfeld (1992) show that bundling is profitable if customers have heterogeneous demand. This is the case in the machine tool industry where customers tend to have very different requirements with regard to the machines they are buying.

Major contributions to the bundling and bundle pricing problems have been made by Hanson and Martin (1990), who formulate a disjunctive mixed integer linear program. The company's profit is maximized subject to constraints that guarantee that each customer's surplus is maximized. In addition to the Hanson-Martin model, we will incorporate fixed costs that are incurred in the production process through the pre-designed modules. Fuerderer et al. (1996) and Fuerderer (1996) extend the Hanson-Martin model by accounting for uncertainty in the customer's choice of a bundle. Both, Hanson and Martin and Fuerderer et al. aggregate customers in broad market segments which seems to be inappropriate for the machine tool industry. Customers demand very different machines and, compared, for example, to the automobile industry, sales are relatively small in numbers. Also, machine tool manufacturers face a very volatile demand due to the fact that they are positioned at the end of the supply chain, for example, see Alexander (1990). Customers will therefore be treated as individuals facing uncertain demand for their own products. In order to hedge against this demand uncertainty, customers are willing to purchase machine tool functionality despite of a lack of immediate need for it. This functionality "banking" or "hedging" can be utilized by the machine tool manufacturer through bundling modules to machines.

4 The Customized Machine Tool Model

In what follows, the customized machine tool problem is presented as a sequence of four decision problems. Each can be formulated as mixed-integer linear program. However, optimizing these problems individually can lead to inefficient solutions for the manufacturer. Alternatively, a general and comprehensive "Customized Machine Tool Model" (CMTM) is formulated. It integrates the four decision problems into a single model. In what follows, the individual models are described in more detail, see also Huchzermeier and Tönshoff (1996).

Excluded from the model is the manufacturer's decision which functions he is going to offer. On the one hand, this is a technical question since a provider of drills does not easily change his production programm to 5-axis mills. On the other hand, the customers' functional requirements indicate which functions to offer. Here, we consider the set of offered functions to be given. A frequently deployed method for the design of new products is *conjoint analysis*. Green and Srinivasan (1978, 1990) provide a comprehensive overview of the relevant literature, techniques, reliability and applications of this marketing method. It can be used in order to gain information about customers' reservation prices for functions. We consider this information as given, too.

First, Module Design. The first decision addressed by the model is the assignment of functions to modules. Here, those candidate modules are determined that are going to be offered to customers. A myopic attitude that seeks ways to lower costs quickly makes companies minimize the design costs only. However, this can lead to suboptimal product designs with respect to sales and attractiveness of the products in the market place. Design costs consist of the costs for designing modules that fulfill certain functions. In addition, a fixed adjustment cost factor is included to capture super- and subadditivity. This is necessary because simple additivity of costs cannot be assumed in practice for module design costs (cf. Section 2.1).

Second, Module Selection. Having determined the optimal module design with respect to design capacity, one needs to combine or to bundle modules to machine tools efficiently. In general, out of all candidate modules, the optimal machine tool needs to be selected for each customer. Each customer is viewed as demanding a certain set of functions. This implies that at least all required functions have to be present in the offered machine. In addition, it is assumed that only one machine per customer is going to be purchased. A customer who wishes to purchase more than one machine is represented as multiple customers.

If many functions are integrated into a rather small number of modules, the manufacturing time required will increase because more complex parts need to be manufactured. However, the assembly time will decrease since less connections between modules have to be made. The fixed costs related to machines account mainly for investments in machinery and equipment and the costs for specialized work force. Super- or subadditive manufacturing and assembling costs similar to the design costs are captured by positive or negative fixed costs, respectively.

Third, Bundle Pricing. The machine tool manufacturer needs to offer machines at prices which maximize his profit allowing customers to obtain only positive consumer surpluses. The next decision is to assign prices to machines that optimize the company′s profit. In the model, reservation prices for machines rather than for functions are used. The reason for this is that the reservation price for functions will differ depending on which functions are contained in a machine. For instance, most customers need a cooling system in their machine tool, thus the reservation price for the cooling system will be considerable. On the other hand, hardly any customer is going to buy a stand-alone cooling system, even if it could be used as an addition to an existing machine. Thus, the reservation price for a cooling system changes with other functions being contained in a machine.

Fourth, Bundle Pricing Under Demand Uncertainty. Depending on the demand a customer expects he will place different reservation prices on machine tools. For example, if demand uncertainty is low, he is going to purchase dedicated machinery. However, if demand uncertainty is high, he prefers a more flexible machine tool that can easily be reconfigured. Reservation prices are assumed scenario-dependent, where each scenario reflects different demand realizations, unlike in the Hanson-Martin model. The machine tool manufacturer′s decision is now to set prices for machine tools that pertain to all scenarios and that maximize total profit. The profit achieved in each szenario is weighted with the probability of realization of the particular demand scenario. The fixed design, manufacturing and assembly costs, however, are incurred independently of scenarios, since the module design and manufacturing activities are decided on before actual demand realizations are revealed.

In order to find a global optimum to the proposed decision problems, one needs to aggregate all of them into one master problem. The strucure of this problem is inherently non-linear, since it includes a quadratic objective function and quadratic constraints. Reformulating the problem using additional auxiliary variables and disjunctive programming, the Customized Machine Tool Model can be formulated as a mixed-integer linear program, see also Tönshoff (1997). This allows us to find a globally optimal solution determining the optimal module design (combination of functions to modules), optimal module selection (integration of modules to machines) and optimal prices for the machines contingent on demand realizations. Figure 4 exhibits the decision process of the *bundling* and *bundle pricing* strategy. Different information, which is provided as input from the customers, marketing and production is shown. Also, the solution output obtained from the model, i.e., modules offered, machines selected and prices charged, is depicted.

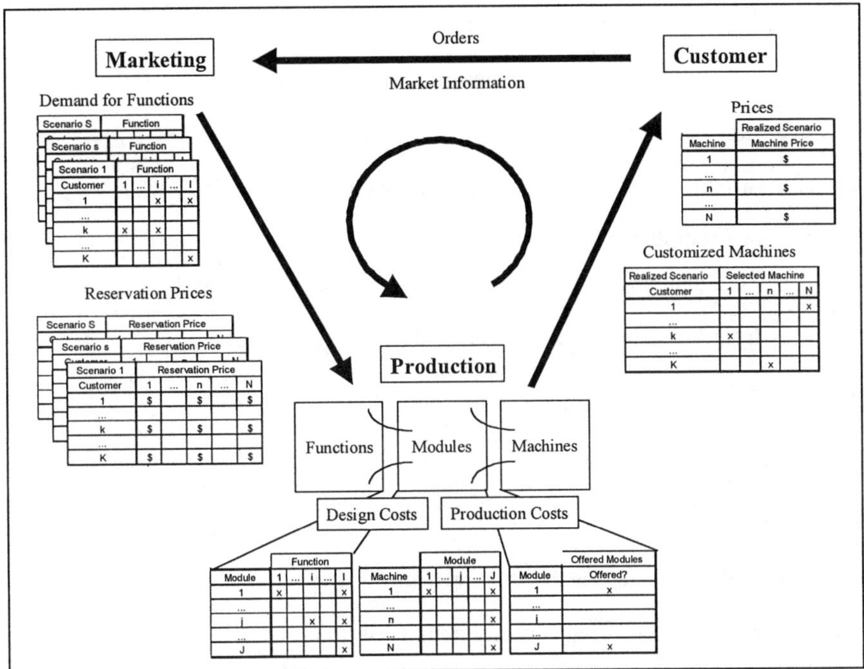

Fig. 4. Decision process of the *bundling* and *bundle pricing* strategy

Indices

i = index of functions, $i = 1,..., I$.

j = index of modules, $j = 1,..., J$.

k, k' = index of customers, $k, k' = 1,..., K$.

n = index of machine tools, $n = 1,..., N$.

s = index of demand scenarios, $s = 1,..., S$.

Problem Parameters

a_{ij} = indicates whether function i is contained in module j or not.

b_{jn} = indicates whether module j is contained in machine tool n or not.

δ_{in} $= \sum_{j=1}^{J} a_{ij} b_{jn}$, indicates how often function i is contained in machine tool n .

$j_{\max}$ = upper bound on the number of offered modules, $(1 \leq j_{\max} \leq J)$.

c_{ij} = variable design costs of including function i in module j (zero otherwise).

c_j $= \sum_{i=1}^{I} c_{ij}$, design costs for module j .

F_j = fixed adjustments to total design costs for module j , $F_j > 0$: superadditivity, $F_j < 0$: subadditivity.

c_{jn} = variable manufacturing and assembly costs for module j when incorporated in machine tool n (zero otherwise).

F_n = fixed manufacturing and assembly costs for machine tool n .

t_{ij}^d = design time of function i for module j (zero otherwise).

t_j^d $= \sum_{i=1}^{I} t_{ij}^d$, total design time for module j .

t_j^m = manufacturing time for module j .

t_j^a = assembly time for module j .

Cap^d = total design capacity in units of time.

Cap^m = total manufacturing capacity in units of time.

Cap^a = total assembly capacity in units of time.

π_s = probability of scenario s , $\sum_{s=1}^{S} \pi_s = 1$.

d_{iks} = demand of customer k for function i under scenario s .

R_{kns} = reservation price of customer k for machine tool n under scenario s .

Binary Decision Variables

y_j = indicates whether module j is offered (zero otherwise).

y_{kns} = indicates whether customer k selects machine tool n under scenario s .

Decision Variables

p_n = the price charged by the manufacturer for machine tool n .

Auxiliary Variables

p_{kns} = price customer k pays for selecting machine tool n under scenario s.

y_{ns} = indicates whether machine tool n is selected for a customer under scenario s.

Objective function

$$\max \sum_{s=1}^{S} \pi_s \left\{ \sum_{k=1}^{K} \sum_{n=1}^{N} \left(p_{kns} - \sum_{j=1}^{J} c_{jn} y_{kns} \right) - \sum_{n=1}^{N} F_n y_{ns} \right\} - \sum_{j=1}^{J} (c_j + F_j) y_i \tag{1}$$

All functions offered?

$$\sum_{j=1}^{J} a_{ij} y_j \geq 1 \qquad \forall i \tag{2}$$

Number of offered modules

$$\sum_{j=1}^{J} y_i \leq j_{\max} \tag{3}$$

Machine selection

$$y_{kns} \leq y_{ns} \qquad \forall k; \forall n; \forall s \tag{4}$$

Customers´ requirements

$$\sum_{n=1}^{N} \delta_{in} y_{kns} \geq d_{iks} \qquad \forall i; \forall k; \forall s \tag{5}$$

All modules offered

$$y_{kns} \leq 1 - b_{jn} + b_{jn} y_i \qquad \forall j; \forall k; \forall n; \forall s \tag{6}$$

Maximum surplus

$$\sum_{n=1}^{N} (R_{kns} y_{kns} - p_{kns}) \geq R_{kns} - p_n \qquad \forall k; \forall n; \forall s \tag{7}$$

Positive surplus

$$R_{kns} y_{kns} - p_{kns} \geq 0 \qquad \forall k; \forall n; \forall s \tag{8}$$

Single price

$$p_{kns} \geq p_n - \left(\max_{\substack{k=1,\dots,K \\ n=1,\dots,N \\ s=1,\dots,S}} \{R_{kns}\} \right) (1 - y_{kns}) \qquad \forall k; \forall n; \forall s \tag{9}$$

$$p_{kns} \leq p_n \qquad \forall k; \forall n; \forall s \tag{10}$$

Design capacity

$$\sum_{j=1}^{J} y_i t_j^d \leq Cap^d \tag{11}$$

Manufacturing capacity

$$\sum_{k=1}^{K} \sum_{n=1}^{N} \sum_{j=1}^{J} t_j^m b_{jn} y_{kns} \leq Cap^m \qquad \forall s \tag{12}$$

Assembly capacity

$$\sum_{k=1}^{K} \sum_{n=1}^{N} \sum_{j=1}^{J} t_j^a b_{jn} y_{kns} \leq Cap^a \qquad \forall s \tag{13}$$

Single purchase

$$\sum_{n=1}^{N} y_{kns} = 1 \qquad \forall k; \forall s \tag{14}$$

Tightening constraints

$$\sum_{n=1}^{N} (R_{kns} y_{kns} - p_{kns}) \geq \sum_{n=1}^{N} (R_{kns} y_{k'ns} - p_{k'ns}) \qquad \forall k; \forall k'; \forall s \tag{15}$$

Binary and non-negativity constraints

$$
\begin{array}{lll}
y_j & \in \{0,1\} & \forall j \\
y_{kns} & \in \{0,1\} & \forall k; \forall n; \forall s \\
y_{ns} & \in \{0,1\} & \forall n; \forall s \\
p_n & \geq 0 & \forall n \\
p_{kns} & \geq 0 & \forall k; \forall n; \forall s
\end{array}
$$

In the objective function, the profit of the company is expressed as the sum of all prices charged to customers due to their selection of machines minus all incurred costs. The latter consist of the production costs for all modules sold, the fixed costs of offering certain machines and the design costs of offering certain modules. In this model formulation, an auxiliary price variable p_{kns} is introduced. This is necessary to maintain a linear model structure. It is assumed that each customer selects exactly one bundle, i.e., one machine tool.

Constraint (2) ensures that all functions are offered in at least one module within the set of offered modules. The number of offered modules is restricted, see constraint (3). Parameter $j_{\max}$ is determined by the machine tool manufacturer who may decide that the maximum number of available modules should not exceed a certain limit. The „Machine selection criterion" ensures that fixed manufacturing and assembly costs are only incurred when a machine is selected by at least one customer. Constraint (5) makes sure that the customers′ requirements for functions are met.

Constraint (6) ensures that all modules needed for a particular machine are being offered. Putting it differently, a customer selecting a particular machine tool forces certain modules to be offered, thus incurring the fixed and variable costs related to the design of the corresponding modules.

Constraints (7)-(10) and (14) set appropriate prices for all machines that allow for a maximum, positive consumer surplus while each customer selects only one machine and all customers pay the same price for the same machine. Constraint (14) forces every customer to select exactly one machine.

For each customer k only one decision variable y_{kns} can take on the value one under each scenario. Consequently, constraint (8) forces p_{kns} to be zero for all other machines for customer k under scenario s. Also, the left hand side of constraint (7) represents the consumer surplus for the one machine that has been selected. This surplus is being compared to all other surpluses the customer could have obtained by selecting different machines. Thus, (7) guarantees that the maximum consumer surplus is gained by each customer. In other words, if a customer has selected a machine tool, then this must be his best choice. The price p_n is set accordingly.

The single price constraints (9) and (10) together with the single purchase constraint (14), render explicit price discrimination between customers impossible so that all customers buying a machine tool are going to pay the same price. All p_{kns} are forced to a value of zero, except for the price of the machine selected by customer k. Hence, constraint (9) ensures that the prices p_n for all machines *not* selected by customer k are at most as high as the maximum reservation price (over all customers, machines and scenarios).

Constraints (11)-(13) prohibit the offering and, as a result, the selection of machines requiring more than the available capacity in design time, manufacturing time and assembly time, respectively.

Finally, the tightening constraint (15) is introduced only to restrict the feasible region of the relaxed mixed-integer program. For a further discussion of the solution methods and possible extensions and reformulations of the model, see Tönshoff (1997).

5 Application Case Studies

In this section, two major German machine tool manufacturers will be described with respect to their bundle design and pricing strategies. Both companies are competing in the global machine tool market (about 60% of sales are being exported). Also, they have a reputation for delivering highly complex machines that satisfy customers with rather difficult production requirements. Both companies were severely hit by the recession in the mid-nineties so that employment figures halved within 3 years.

5.1 The Opportunistic Selling Company

Company A produces lathes and CNC-lathes in Germany and in various other countries in Europe, America and Asia. Virtually all of the machines contain some parts that are customer-specific. In many cases, these parts have to be designed from scratch. In order to capitalize on scale economies, subassemblies are produced regardless of customer orders. When a customer order arrives, the necessary subassemblies are taken from inventory and customer-specific items are being manufactured.

Decisions on how many of each subassembly have to be produced are based on a master forecast of the number of machines or machine types that are going to be sold during the planning horizon. Each machine requires a list of subassemblies that are mandatory or optional. It is "vital" for this system to have good estimates for machine sales as well as for the optional parts.

At company A, the bundle strategy is known as offering *Packaged Machines* and has been used for a couple of years. However, only about 10% of all sales are Packaged Machines, whereas 90% are still customer-specific machines with long lead times.

This small share of sales is due to the fact that offering Packaged Machines is used mainly as a marketing incentive, when sales are not very high and capacities are underutilized. Marketing proposes types of machines that are expected to sell well. Also, production suggests machines that should be produced in order to fully utilize the manufacturing capacities. After an agreement is reached, a number of machines with a fixed range of options is produced. A comparatively large volume of a certain machine including subassemblies is manufactured. This is why simpler production planning and control as well as economies of scale allow the company to offer these Packaged Machines at a low price. By this, sales can be stimulated.

Talking to both the production and the marketing department one gets the impression that neither one is content with the effects of the Packaged Machines. One of the main selling arguments is the low price. Customers tend to accept the offer but ask for changes of the Packaged Machine before it is delivered. Some options are requested to be left out and, consequently, the overall price of the machine is expected to be lowered even further. Alternatively, some other options should be included. The sales engineer is afraid of the customer not buying the machine and agrees to the required changes. It is not considered, that these changes do not only incur costs for adding options but also for taking out options and for loosing the anticipated scale effects. Thus, the production department has to change the machine, sometimes even after it has been through final assembly, and production costs are increased substantially. This leads to complaints about the marketing department not complying to agreements. Also, since customers are often not willing to accept the Packaged Machine without change, sales personnel is unhappy with this strategy. It was not possible to avoid these problems, even though historical demand data was used to design the Packaged Machines.

In the case of company A, "Packaged Machines", i.e., module bundles, were to be proposed that could be sold for optimal prices in certain customer segments. For this, the modules were assumed to be given.

Since not all of the data supported by the CMTM model was available, the model was adapted to the data structure found at company A. Furthermore, we concentrated on one type of machine only. Production and assembly costs were split up and allotted to modules. Fixed costs were not available. Only some of the assembly time requirements were included. Thus, capacity restrictions did not constrain the solution space.

The following four major markets were split into eleven submarkets. "Typical customers" for these markets were determined. Each customer had a certain demand structure, i.e., required a number of modules and had different reservation prices for modules.

- Automotive suppliers from Northern Europe (Scandinavia and the UK)
- Automotive producers from Northern Europe (Scandinavia and the UK)
- Automotive suppliers from the US
- Automotive suppliers from Germany

About 20 different major options of the machine were considered. Excluding modules that were almost never required by any customer and including modules that are always required into the "basic machine", finally left 10 modules.

Using GAMS™ (General Algebraic Modeling System) the Customized Machine Tool Model was generated and compiled. The GAMS/CPLEX solver was then used to solve the problem to optimality on a 133 MHz Pentium machine with 48 MB RAM. Not using the tightening constraints, the model included more than 50,000 single equations and more than 30,000 single variables. GAMS needed 0.43 seconds for the syntactical check of the model. The generation time that was required for preparing the model for solution was 77.61 seconds. CPLEX did 24,245 iterations that took 2,363.99 seconds to solve the mixed-integer program and another 6 iterations to solve the final LP. Finally, it took GAMS another 15.66 seconds to read and output the results. The relative gap between the best integer solution possible and the final solution was zero, making the final optimal solution the same as the best integer solution possible. The Customized Machine Tool case study model including the extensions of a *minimum profit constraint* and a *null-bundle* was also solved to optimality. The optimal solution was identical to the one without these extensions.

The results include a proposal for a machine tool for each submarket and an *optimal* price for this machine. Looking at the single customers one sees that all customers get at least all the required modules. In case a customer is offered a machine that contains more than the required modules, it "pays" for the manufacturer. The *optimal* prices are sometimes lower, sometimes higher than the *list* prices that the manufacturer offers the machines for. The manufacturer admitted that high discounts are almost always given to customers. One reason is that for some modules the reservation prices are lower than the *list* prices. Thus, in order to sell the machines to customers, the manufacturer needs to lower his *list* prices. In the past, this led to Packaged Machines not making any profit. Comparing the profit obtained by offering the machines at the *optimal* prices (as given in the optimal solution) and at the *list* prices, profits could not be increased (a minus of about 5%). Comparing the *optimal* prices with the past *realized* prices (including the discount), the profit could in fact be increased by about 3%.

5.2 The Fast Innovator Company

Company B is a major producer of lathes, CNC-lathes and turning automatics. Offering simple machines in addition to their sophisticated production equipment has been suggested and exercised repeatedly in the company's history. Simple machines were built and offered to customers. However, whenever this approach was followed, customers complained that these simple machines do not provide all the functionality needed. In response, options and additional functions had to be offered. Technical inventions and developments provided the company with a competitive advantage that took competitors some time to copy and incorporate into their products.

For a couple of years, company B has been offering machines that are configured using a modular system. The customer chooses the basic machine type based, e.g., on the size of parts he wants to manufacture, and adds options according to his necessities. He might, for instance, choose a single spindle or two synchronized spindles. There exists only one design for a spindle. Thus, not only are the main spindles in both alternatives the same, but also is the synchronized spindle in the second alternative the same as the main spindle. This standardization leads to comparatively high scale effects. Since all machines get the same spindle, its design may be "over-engineered" for some customers. However, designing and producing different spindles and choosing the appropriate one for each customer is assumed to lead to higher costs without enhancing customer satisfaction.

Offering these modular machine tools lets the customer decide what he needs and enables company B to provide an almost totally customized product at a competitive price. Customers who require basic machines and little options only can be satisfied as well as those who need a highly complex production task to be solved. In reality, however, only very few of the simple machines that could also be thought of as potential volume products are sold. In fact, about 5% of the sales are made with simple machines and most sales belong to the high-end segment.

Reasons for this could be that the highly sophisticated machines are being sold at prices that are competitive or that company B has a monopoly on these machines. On the other hand, the low-end machines may still not be offered at a competitive price. This is possibly due to the fact that the cost structure of a German machine tool builder is still very disadvantageous in the international comparison. It is also possible that German machine tool manufacturers do not fulfill customers' expectations if they sell low-end machinery.

The fact that only 5% of the sales are simple machines suggests that there is room for a better configuration of the "catalogue" of options that customers choose from. If many of the options are almost always demanded then some of them could be combined into a module or even added to the basic machine layout. This would allow for higher standardization and cost savings. Also, this facts supports the hypothesis that the low-end machines are too expensive compared with foreign competitors. Thus, even using standardized modules does not enable a German manufacturer to compete on costs/price successfully against low-wage competitors for certain products. Our analysis suggests that a trade-up policy might generate more sales in the low-end segment of the market.

The throughput time of customer-ordered machines has been decreased substantially (by about 75%) by means of modular design. This implies that the manufacture and assembly of customer-specific parts generally is the bottleneck in the process of manufacturing and assembling a machine. If these parts could be anticipated to some extent and, as a result, standardized and preproduced, the throughput time could be lowered even further.

The cases have shown that both companies are aware of the bundling of modules to machines, but only company B has already standardized many of the subassemblies in order to gain scale effects. Company A so far only uses bundles as a marketing tool to increase sales and to raise the capacity utilization. Company B has introduced a complete modular machine line which is marketed aggressively. Company B is not using its advantages of being able to provide customer-specific solutions at relatively low prices to increase profits by demanding price premia.

Both face the problem that customers are complaining about German "over-engineering" but when offered simpler and cheaper machines they ask for more options and sophistication. If additional options are not offered the customers threaten to buy a standard Japanese machine.

6 Implementation and Solution

The Customized Machine Tool Model (see section 4) with the case study data (see section 5) was solved to optimality using the *GAMS* system with the *CPLEX* solver. Computational effort and solution times are reported in section 5.1. It is interesting to learn that none of the other solvers was able to solve this problem. *Excel* and *LINGO* did not allow problems of this size. *GAMS* with the different solvers that could be used had special problems as well. Thus, only the *CPLEX* solver was able to find the optimal solution. Before this was possible, some additional program reformulations needed to be implemented. First, the maximum reservation price that is used in the "Single price" constraint (9) is evaluated before the optimization starts. This is done by introducing an additional auxiliary parameter which is defined as maximum reservation price. In constraint (9) this parameter is used. Second, the "Maximum consumer surplus" constraint (7) causes *GAMS* to produce a memory overflow error when generating the internal model representation. This can be circumvented by splitting the constraint into three parts and introducing auxiliary variables that are set equal to these parts of the original equation. The "Maximum consumer surplus" constraint then contains these auxiliary variables only. The authors thank Alexander Meeraus from *GAMS* Development Co. for benchmarking the proposed reformulation and confirming the results. Only with these implementations, *GAMS* was able to generate the model that *CPLEX* then solved to optimality.

7 Conclusions

In this chapter, two leading companies in the German machine tool industry were analyzed. A stochastic optimization model formulation has been proposed for the customized machine tool module design, module selection and bundle pricing problem. The model was applied to real-world data and the results were benchmarked against the industry practice. One of the results is that German machine tool manufacturers, who are able to provide customized machine tools using a modular product line, are not able to employ their advantages to increase profits by demanding price premia. Besides the disadvantages of the cost structure of a German machine tool builder this is possibly due to disregarding reservation prices and therefore customers' needs.

It has been shown how the standardization and modularization strategies can be extended with a bundling and bundle pricing strategy. The potential impact on a manufacturer's profitability and its competitiveness in production planning became evident. By considering customers' reservation prices and optimizing the offered modular product line the "gap" between marketing and manufacturing can be narrowed if not closed.

References

Alexander, A. J. (1990). "Adaptation to Change in the U.S. Machine Tool Industry and the Effects of Government Policy." RAND Note N-3079-USJF/RC, RAND Corp., Santa Monica, 20-21.

Blackburn J. D. (1991). "Time-Based Competition." Irwin, Homewood.

Dobson, G. and S. Kalish (1993). "Heuristics for Pricing and Positioning a Product-line Using Conjoint and Cost Data." Management Science, Vol. 39, 2 (February), 160 - 175.

Eppen, G. D., W. A. Hanson and R. K. Martin (1991). "Bundling-New Products, New Markets, Low Risk." Sloan Management Review, Vol. 32, 4 (Summer), 7-14.

Finegold, D., K. W. Brendley, R. Lempert, D. Henry, P. Cannon, B. Boultinghouse and M. Nelson (1994a). "The Decline of the U. S. Machine-Tool Industry and Prospects for Its Sustainable Recovery, Volume 2." Appendices, Critical Technologies Institute RAND, Santa Monica.

Finegold, D., K. W. Brendley, R. Lempert, D. Henry, P. Cannon, B. Boultinghouse and M. Nelson (1994b). "The Decline of the U. S. Machine-Tool Industry and Prospects for Its Sustainable Recovery, Volume 1." Critical Technologies Institute RAND, Santa Monica.

Fisher, M.L. (1994). "National Bicycle Industrial Co." Case Study, The Wharton School, University of Pennsylvania, Philadelphia.

Franke, H.-J., A. Jeschke and H. Speckhahn. (1995). "Standardisierung komplexer Baugruppen." ZwF, Vol. 90, 1-2, 46-48.

Fuerderer, R. (1996). "Option and Component Bundling Under Demand Risk." Deutscher Universitätsverlag, Wiesbaden.

Fuerderer, R., A. Huchzermeier and L. Schrage (1996). "Stochastic Option Bundling and Bundle Pricing." Working paper 94-12, WHU-Otto-Beisheim-Hochschule, Vallendar (April).

Gabel, H. L. (1991). "Competitive Strategies for Product Standards." McGraw-Hill, Maidenhead.

Green, P. E. and A. M. Krieger (1985). "Models and Heuristics for Product Line Selection." Marketing Science, Vol. 4, 1 (Winter), 1-19.

Green, P. E. and A. M. Krieger (1991). "Product Design Strategies for Target-Market Positioning." Journal of Product Innovation Management, Vol. 8, 189-202.

Green, P. E. and V. Srinivasan (1978). "Conjoint Analysis in Consumer Research: Issues and Outlook." Journal of Consumer Research, Vol. 5, (September), 103-123.

Green, P. E. and V. Srinivasan (1990). "Conjoint Analysis in Marketing: New Developments With Implications for Research and Practice." Journal of Marketing, Vol. 54, 4 (October), 3-19.

Hamel, G. (1996). "Strategy as Revolution." Harvard Business Review, Vol. 74, 4 (July-August), 69-82.

Hanson, W. and R. K. Martin (1990). "Optimal Bundle Pricing." Management Science, Vol. 36, 2, 155-174.

Hardle, B., L. M. Lodish, J. Kilmer, D. R. Beatty, P. W. Farris, A. L. Biel, L. S. Wicke, J. B. Balson and D. A. Aaker (1994). "When Does Variety Become Redundancy?", Harvard Business Review, Vol. 72, (November-December), 53-62.

Huchzermeier, A. and N. Tönshoff (1996). "Schneller, kosteneffizienter, profitabler: Kundenorientierte Produktdesignstrategie im Werkzeugmaschinen-bau." VDI-Z, Vol. 138, 11/12 (Nov./Dec.), 54-57.

Jorissen, H. D., K. Malle and H. J. Schulte (1993). "Produktionstechnik in Erwartung der Konjunkturwende." VDI-Z, Vol. 135, 11/12, 19-22.

Jovane, F. (1989). "The Elementary Machine: An 'Atomic' Model to Analyze and Devise Production Systems." Annotations of the CIRP, Vol. 38, 1.

Kekre S. and K. Srinivasan. (1990). "Broader Product Line: A Necessity to Achieve Success?" Management Science, Vol. 36, 10 (October), 1216-1231.

Kohli, R. and R. Sukumar (1990). "Heuristics for Product-Line Design Using Conjoint Analysis." Management Science, Vol. 36, 12 (December), 1464-1478.

Kotha, S. (1994). "Mass Customization: A Source of Competitive Advantage or Competitive Suicide?" Working paper Leonard Stern School of Business, New York University, April.

Kotha, S. and A. Fried (1993). "The National Bicycle Industrial Company: Implementing a Strategy of Mass-Customization." Case study, International University of Japan, May.

Lee, H. L., C. Billington and B. Carter (1993). "Hewlett-Packard Gains Control of Inventory and Service through Design for Localization." Interfaces, Vol. 23, 4 (July-August), 1-11.

Lele, M. M. (1992). Creating Strategic Leverage. Wiley, New York.

Lilien, G. L., P. Kotler and K. S. Moorthy (1992). Marketing Models. Prentice-Hall International, Englewood Cliffs.

Malle, K. (1994). "Aufgabenorientierte Standard-Drehmaschinen: anpaßbar, weil modular." VDI-Z, Vol. 136, 4, 68-70.

McBride, R. D. and F. S. Zufryden (1988). "An Integer Programming Approach to the Optimal Product Line Selection Problem." Marketing Science, Vol. 7, 2 (Spring), 126-140.

McCutcheon, D. M., A. S. Raturi and J. R. Meredith (1994). "The Customization-Responsiveness Squeeze." Sloan Management Review, Vol. 36, Winter, 89-99.

Pahl, G. and W. Beitz (1993). Konstruktionslehre: Methoden und Anwendung. Springer, Berlin, Heidelberg, New York, 3rd ed.

Pahl, G. and W. Beitz (1984). Engineering Design. K. Wallace (Ed.), Springer, Berlin, Heidelberg, New York.

Pindyck, R. S. and D. L. Rubinfeld (1992). Microeconomics. Macmillan, New York.

Pine, B. J. II. (1993). "Mass Customization: The New Frontier in Business Competition." Harvard Business School Press, Cambridge.

Quelch, J. A. and D. Kenny (1994). "Extend Profits, Not Product Lines." Harvard Business Review, Vol. 72, (September-October), 153-160.

Reiß, M. and T. C. Beck (1994). "Fertigung jenseits des Kosten-Flexibilitäts-Dilemmas." VDI-Z, Vol. 136, 11/12, 28-30.

Roever, M. (1992). "Curing the Disease of Overcomplexity." The McKinsey Quarterly, Vol. 2, 97-104.

Salinger, M. A. (1995). "A Graphical Analysis of Bundling." Journal of Business, Vol. 68, 1, 85-98.
Siemens-Zeitschrift "Werkzeugmaschinenbau im Umbruch." Siemens-Zeitschrift, Vol. 6 (1995), 20-24.

Spur, G. (1996). "Perspektiven des deutschen Werkzeugmaschinenbaus." ZWF, Vol. 91, 6, 244-245.

Stalk, G. (1988). "Time - The Next Source of Competitive Advantage." Harvard Business Review, Vol. 66, 4 (July-August), 41-51.

Stalk, G. and T. M. Hout (1990). "Chapter 9: Time-based Strategy." In: Competing Against Time, The Free Press, New York.

Starr, M.K. (1965). "Modular Production - A New Concept." Harvard Business Review, Vol. 43, (November-December), 131-142.

Stigler, G. J. (1963). "United States v. Loew's Inc.: A Note On Block-Booking." The Supreme Court Review, 152-157.

Tönshoff, H. K. and F. Böger (1994). "Entwicklung einer Referenzarchitektur für den modularen Aufbau zukünftiger Werkzeugmaschinen." wt-Produktion und Management, 84, 330-333.

Tönshoff, N. (1997). Modular Machine Tools - Bundling and Pricing Strategies Under Demand Uncertainty. Gabler, Wiesbaden.

Ulrich, K. T. and K. Tung (1991). "Fundamentals of Product Modularity." Working paper MIT Sloan School WP No. 3335-91-MSA (September).

Ulrich, K. T. and S. D. Eppinger (1995). Product Design and Development. McGraw-Hill, New York.

Ulrich, K. T., D. Sartorius, S. A. Pearson and M. Jakiela (1993). "Including the Value of Time in Design-for-Manufacturing Decision Making." Management Science, Vol. 39, 4, 429-447.

VDMA (1994a) (Verband Deutscher Maschinen- und Anlagenbau e.V.). "Kennzahlenkompaß." VDMA, Frankfurt.

VDMA (1991) (Verband Deutscher Maschinen- und Anlagenbau e.V.). "Kosten-Kennzahlen im Machinen- und Anlagenbau 1990." VDMA, Frankfurt, December.

VDMA (1993b) (Verband Deutscher Maschinen- und Anlagenbau e.V.). "Statistisches Handbuch für den Maschinenbau 1993." VDMA, Frankfurt.

VDMA (1994b) (Verband Deutscher Maschinen- und Anlagenbau e.V.). "Statistisches Handbuch für den Maschinenbau 1994." VDMA, Frankfurt.

VDMA (1995) (Verband Deutscher Maschinen- und Anlagenbau e.V.). "Statistisches Handbuch für den Maschinenbau 1995." VDMA, Frankfurt.

Westbrook, R. and P. Williamson (1993). "Mass Customization: Japan's New Frontier." European Management Journal, Vol. 11, 1, 38-45.

Womack, J. P. (1993). "Book Review on: Mass Customization." Sloan Management Review, Vol. 35, Spring, 121-122.

Market-Oriented Complexity Management Using the Micromarket Management Concept

Jürgen Ringbeck[1], Carl-Stefan Neumann[2], and Andreas Cornet[3]

[1] **Jürgen Ringbeck**, McKinsey & Company, Inc., Duesseldorf, Germany.
[2] **Carl-Stefan Neumann**, McKinsey & Company, Inc., Frankfurt, Germany.
[3] **Andreas Cornet** McKinsey & Company, Inc., Duesseldorf, Germany.

1 Introduction

"You can get our Ford model T in any color -- as long as it's black." This was Henry Ford's formula for success in the early days of mass production. Back then, affordable prices for products offering a new level of convenience more than made up for a lack of variety.

The trend in mass production today is marked by a sharp increase in product variety, especially in mature markets, where competitive pressure keeps prices down and quality up, and new products are being launched at shorter and shorter intervals. From 1980 to 1994, for example, Jacobs-Suchard quadrupled the number of its purple cow "Milka" candy products. Last year, the food manufacturer Nestlé introduced 135 new products. In 1995, the food industry introduced a total of 1,200 new products.

In the automotive industry, product cycles shrank from eight to four years between 1980 and 1995. Almost simultaneously, the number of market segments in the US passenger car market increased from 9 to 12. The increasing individualization of standardized mass-market products has motivated Mercedes-Benz and Swatch to plan a series of boutiques in department stores, train stations, and airports to sell their new micro-compact car. In each boutique, customers will use a computer terminal to design and order their own personalized Swatch-mobile by choosing from a multitude of color and equipment options.

In any industry, customization of product offerings entails the rigorous application of three areas of know-how: the principle of modularity in production, bundling strategies in marketing, and information technology (IT) in product and portfolio design. This paper focuses on IT and the role it plays in managing complexity and developing bundling strategies. We use two examples to illustrate the details.

Most companies have plenty of data on customer behavior and preferences to which this know-how can be applied to create new products. The real problem is knowing when or where to stop. At what point does targeting a product line to smaller and smaller markets -- micromarkets -- no longer pay off? Classic cost accounting systems do not accurately allocate complexity costs, the inevitable result of product variety, to specific products. For this reason, customization strategies can easily turn out to be more costly than profitable.

To address this problem, McKinsey has developed a concept for successful micromarket management. It shows ways to use more individualized products to increase the benefit perceived by customers and, at the same time, boost profits. It makes complexity manageable and turns product differentiation into a profitable strategy.

Yet applying this concept is a major challenge even for successful companies. Micromarket management requires the thorough reorganization of existing systems for selecting and processing information and the fundamental restructuring of product planning processes. In our experience, successful implementation of micromarket strategies involves adjustments in four main areas (Figure 1):

- The company's product line must be adapted to the increasing fragmentation of customer value. In the extreme case, the company will need to develop mass customization strategies to meet the requirements of the often cited "segment of one."
- The entire organization must be set up for a much shorter product planning cycle, with far-reaching consequences for product development and supply-chain management.
- Opportunities to reduce costs through synergies in supply must be exploited systematically in product development, production, and distribution.

- On the demand side, superior coverage of many different micromarkets using a common set of (limited) resources must be ensured by carefully defining and matching customer segments with applications and sales channels in order to obtain maximum benefit from synergies.

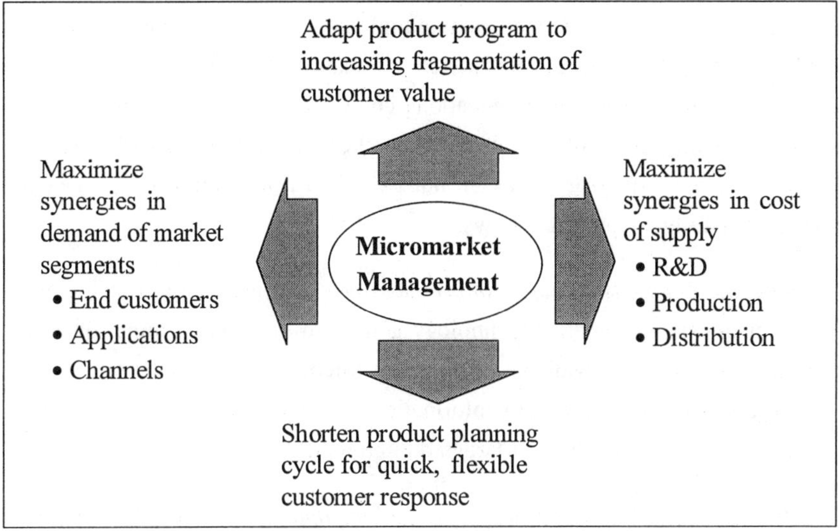

Fig. 1. Key adjustments for successful micromarket management

With regard to the last two points, approaches based on the bundling of products and options have proved very successful. In the automotive industry, for example, it is now common practice to offer optional features only in bundles rather than as separate items and to deliver certain accessories (for example, air conditioning systems) as standard equipment. The effect is twofold. First, careful pricing of option bundles "seduces" customers into paying more for extras along with buying some less popular features as part of the package. Second, manufacturing productivity also improves because supplying options in bundles reduces the total number of possible model variants and increases the unit volume of features that were previously less popular. This leads to cost savings due to economies of scale. In the early 1980s, Chrysler was able to lower the prices for its Dodge Omni and Plymouth Horizon by more than $1,000 with a bundling strategy. At the same time, it reduced the number of option combinations from 8 million to 42.

Efforts to master micromarket management will pay off for companies that sell very complex product lines in highly differentiated markets and are under great pressure to keep changing their products. Such pressure may result from radically shortened product life cycles, from the superior innovativeness of a particular competitor, or from deregulation of the marketplace. Industries currently experiencing such pressures include the machine tool and automotive industries, service providers such as hospitals and insurance companies, and network-based industries such as railroads, telecommunications companies, and airlines. In these industries, the management of micromarkets has become an integral part of suppliers' growth and survival strategies. In our experience micromarket management can be successfully adopted as follows:

- *Define micromarkets and optimal product lines:* Starting with a strategic business diagnosis, information technology is used to evaluate customer data (for the most part already available), the various micromarkets, and the actual costs of product differentiation. This information is then reconciled into a definition of the optimal product line for targeted micromarkets.
- *Define the organizational changes needed for optimal product planning:* The next step is to determine the future requirements for the processes and structures needed for optimal product line planning.
- *Convert products and planning processes in coordinated migration:* The resulting concept for changing the product line and the organization of the company is then implemented with controlled migration management.

2 Defining the Optimal Product Line: Using Information Technology Achieves Customer and Cost Transparency

Uncertainty about the impact of a product line on profits rarely results from a lack of data. More often, companies misinterpret the information they have. Very often, the reason is poor coordination between product development, manufacturing, or marketing and sales. Another reason is that the wealth of additional information resulting from the increasing interaction of manufacturers, the trade, and customers is not yet used systematically for planning and control. Superior information management is indispensable for serving micromarkets profitably. The solutions found by Lufthansa as an international service provider and by a major European components manufacturer illustrate how the concept is applied and what results it can deliver.

2.1 Lufthansa's Success: Learning from Simulated Test Customers in Virtual Micromarkets

Lufthansa offers a successful example of how a company can learn from simulating customer responses to product changes in virtual micromarkets and how the results can be used to optimize the products it actually offers. The first step for Lufthansa was a strategic diagnosis of its product offerings. Next, data on all relevant market trends, competitors' products, and the needs of Lufthansa's own customers were linked electronically using the latest data processing methods. Then Lufthansa's own flight program was customized to better meet the needs of its key customer segments. As a result of this micromarket management approach, profits improved substantially.

In 1992, all the major airlines were still suffering heavy losses. The abolition of exclusive operating rights and the deregulation of prices for flights within Europe initiated fierce competition among airlines. They competed for passengers with a wide array of discounts, special rates, and frequent-flyer offers. The number of different fares offered on major European routes increased by a factor of more than six. Transport capacities were expanded by more than one third.

Faced with this situation, Lufthansa undertook a major restructuring of its passenger business. The overall aim was to adapt product planning to micromarkets. Each segment of air travel between the approximately 200 cities served by Lufthansa has its own unique characteristics - competition, market size, customer preferences, travel times - and for every single flight, the customer makes a new purchase decision. These are typical micromarkets; Lufthansa has more than 20,000. Previously, product planning had been centered on approximately 2,000 large markets, mainly served by direct flights. Now, the remaining 90% of the micromarkets are being integrated into the network planning system.

The first step in the analysis was to combine the data from more than 1 million weekly direct and connecting flights between 200 arrival and departure airports with the passenger information on more than 10 million air travel tickets. This produced a detailed picture of passenger movements and the utilization of individual connections. In addition, the offerings of all of Lufthansa's competitors were evaluated. Finally, several thousand interviews were conducted with passengers. The result was a clear profile of Lufthansa's strengths and weaknesses as well as insight into the airline's most important customer segments and their preferences.

Simulated "test-marketing" using specially developed software was the next step to determine the best way for Lufthansa to make its flight offerings more attractive than those of competitors. For the simulation, data were selected on a number of homogeneous passenger groups. The proprietary software then calculated the "virtual" responses of these passengers to each possible Lufthansa flight compared with the flights offered by competitors. The biggest source of complications in flight planning is that most flights function both as nonstop flights and as transfer connections in a package of flights. Passenger preferences differ accordingly. While direct flyers want attractive arrival and departure times, passengers who change planes want short transfer times in order to shorten their total travel time. The purpose of optimization was to make the flight connections more attractive, both as nonstop flights as well as flights within a package of other flights. Therefore, both preferences had to be taken into account.

Before micromarket management was introduced, the flight plan was set up by regional route managers who mainly concentrated on the demand for nonstop flights and tried to achieve the best possible departure times. In addition to changing this situation, management had to implement the new route plans and schedules with existing equipment and staff.

The new integrated planning approach now combines product design and resource planning. The design of flight connections is based on a "departure-destination concept." Every flight carries passengers with many different departure and destination locations (i.e., micromarkets). The passengers have to transfer from and to different flights during the course of their itinerary. Thus, the optimization of each flight involves multiple micromarkets. A special software had to be developed in order to integrate the various market and product requirements into the flight planning process.

The results speak for themselves. Without expanding aircraft capacity, Lufthansa was able to win more customers. More importantly, they won customers who were more attractive from an economic standpoint. The bottom line was that using micromarket management improved profits by several hundred million marks.

2.2 Computer-Supported Concepts to Achieve Cost and Profit Transparency in Industrial Components Manufacturing

Micromarkets can be penetrated by offering a new flight, a new options package for a passenger car, or an additional function on a machine. But serving each new micromarket also generates additional costs. Therefore, an important challenge of micromarket management is to make the costs of product differentiation transparent.

Why do many companies still experience uncertainty about the actual cost of adding a new product variant to their existing product lines? Generally, overhead costs - which in Germany account for a significant share of total costs - are allocated in a lump sum to all products in job-based cost accounting. Thus, complexity costs, such as those resulting from the need to retool a machine or order special material for a small lot size are systematically undervalued, and the profitability of products produced in small lots tends to be overvalued.

The only way to achieve the necessary transparency in identifying the revenue and cost effects of individual product line decisions is with systematic, IT-supported calculation of actual product costs. By taking into account interdependencies between product lines and production processes, it becomes possible to estimate the full ramifications of changes within the product portfolio as a whole. This understanding is essential to grasp the consequences of complexity on a conceptual level and, as a result, to exert better control over the factors that affect sales and profitability.

A leading German manufacturer of industrial components has demonstrated that it is possible to gain such an understanding of the true costs of product complexity within only a few months. The findings guided the company in redesigning its product lines for more profitability. Like Lufthansa, this company achieved superior results through consistent application of IT as part of a strategic product line diagnosis. Using a multi-stage approach in which computer-supported calculation of the costs and profitability of different product versions played a key role, the company obtained an overview of its actual process costs (Figure 2):

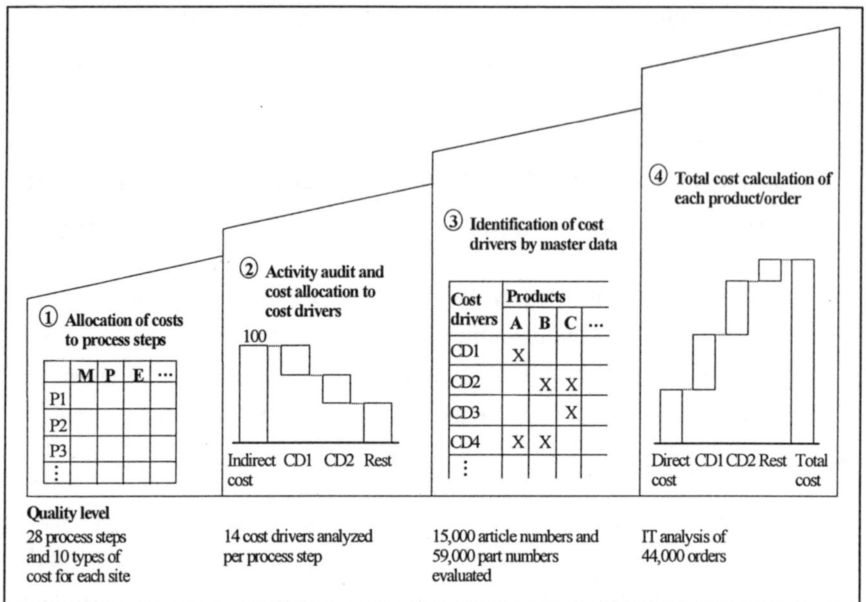

Fig. 2. Introduction of process-oriented cost calculation in 6 to 8 weeks

- The first step was to divide all the areas of the company into process steps, each exhibiting typical complexity characteristics. Individual costs were allocated to these process steps on the basis of operating statistics.

- Management identified the principal complexity drivers for each process step and described the causal connection between product-related complexity and process costs.

- With existing bill-of-materials data, a new product costing database was established. In addition to other complexity-driving factors, ordering behavior was included as an important new factor in the new IT-supported product cost calculation. For this purpose, an IT application was developed to code and analyze more than 44,000 orders.

- Based on the resulting product calculation database, more than 15,000 existing articles and 59,000 existing parts were recalculated.

The payoff of this major analytical effort was a new and more accurate view of the component manufacturer's cost structure. In the past, management had assumed that unit costs could not be reduced significantly as a function of lot size.

Each production run appeared to be equally attractive, and variety seemed to come free of charge. The more accurate calculations made possible with IT made it clear that the actual costs for manufacturing small runs were as much as four times greater than for products produced in large runs (Figure 3). For the first time, the limits of profitable complexity became transparent.

This more accurate understanding of the cost structure and the ability to assess variety-driven costs laid the groundwork for a total redesign of the company's product line. The objective of the redesign was to meet customer requirements with dramatically fewer different parts, thus reducing company-internal complexity.

In the first step, managers reviewed their current and prospective customers with the aim of defining segments of similar customers who could be served by similar products. As a result, they found a way to join two production lines, which reduced product variety by almost half.

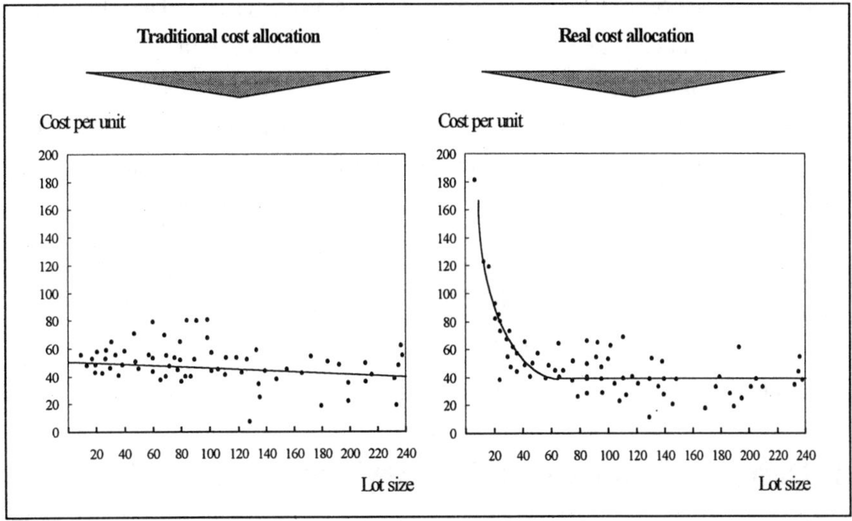

Fig. 3. Scale effects visible only with *process-oriented* allocation of indirect costs

As a second step, the company took a careful look at the parts for each product line and conducted a rigorous standardization program. For several parts, managers were able to reduce variety substantially, even by as much as 90 percent. In some cases, the design of standardized feature bundles led to a slight overdesign of the product for certain applications where the customer does not require all the features contained in the bundle. Although this led to higher direct variable costs, the reduced complexity costs more than compensated for this effect. With their new cost understanding and the ability to assess complexity costs, the company's managers were able to see this cost trade-off and find the optimum. Information technology again played a major role, ultimately enabling the company to achieve an overall reduction of 75% measured in different parts.

Despite such proven successes, many companies are still wary of subjecting their product lines to a strategic diagnosis to identify genuine additional benefits as well as complexity costs, and of using information technology as a critical tool for obtaining greater transparency.

They are worried about the time and expense required to establish sufficiently comprehensive databases. One reason for this is that they are not fully aware of recent advances in IT. Another reason is that they have had bad experiences with large-scale computer projects.

However, successes like those achieved in the airline or manufacturing industries show that an IT-supported analysis of customer needs and of actual complexity costs is often the only way for companies accurately to plan their product offerings with an optimal cost-profit trade-off. Striking a balance between these two objectives is the only way to guarantee sustainable corporate growth and profitability.

Merely tapping new sources of information through a one-time product line planning project is not enough. Micromarket management also requires that the structures for radically redesigned, optimized product line planning must be solidly rooted in the organization.

3 Defining Fundamentally New Processes and Structures for Micromarket Management

Companies are now facing challenges that are a direct result of the micromarket management concept and that go well beyond the narrow scope of projects used to redesign product lines. To stay competitive, top management has to adapt corporate structure to the results of the project in a fundamental and permanent way.

3.1 New Product Line Planning and Information Management Processes

Micromarket management makes innovative use of information technology to interpret available information more precisely for product line planning. It is only natural that there are entirely new demands with regard to corporate information management. Even the computer-aided maintenance and preparation of the planning data is a completely new responsibility. For example, Lufthansa Systems, a subsidiary of Lufthansa, uses a comprehensive data warehousing concept to analyze booking and billing data (more than 20 million passenger bookings are made annually). "Warehousing" is the only way the huge volume of data can be used continuously for flight planning and control purposes.

In the manufacturing sector, the introduction of new computer-supported product calculation tools for sales and marketing, so-called "product configurators", or the introduction of activity-based costing and new simulation tools in product development have required significant changes in information management. At the same time, the use of these analytical tools has led to fundamental changes in the processes employed in product development, sales and marketing. Many coordination and approval processes, which were previously sequential, can be made "parallel" and replaced by cross-functional team solutions. Also, structural mergers of previously separate areas in product development and marketing may be required.

3.2 New Organizational Structures

In the case of the component manufacturer discussed in the previous section, the micromarket management approach led not only to a fundamental redesign of the product line, reprioritization of product development projects, and a completely new computer-supported calculation system, but also to the reorganization of certain areas within manufacturing.

A small Japanese bicycle manufacturer, NBIC, which has become known as an example of successful mass customization, took the process even further. The company's goal was to offer a customized bicycle at an additional cost of only about 20 to 30 percent over standard mass produced bicycles and with a delivery time of about two weeks. After thoroughly assessing the problem, the company decided to separate the customized manufacturing operations completely from the high-volume business. Product planning and organization were specifically redesigned to meet the needs of the specialty business. NBIC is now selling about 12,000 customized bicycles each year.

4 Controlled Introduction Through Migration Management

The changes in processes and structures needed for micromarket management are so fundamental that their implementation must be led by top management. In view of the risks involved - organizational restructuring and, at the same time, significant changes in the product line - it is wise to implement the new product line planning process in two steps:

- *Detailed design and testing of the planning process*: In this stage, the individual steps within the new planning process are defined in greater detail. This detailed design is then verified through pilot tests (for example, in selected regions or product lines). At the same time, the required IT infrastructure for data evaluation and analysis has to be set up - initially in the form of prototypes. The purpose of this first phase is to check specific product line decisions in the marketplace using test cases, thereby ensuring the robustness of the innovative process.

- *Full implementation*: The success of the pilot tests determines whether the product line should be redesigned and, if so, what the nature of the changes will be. Full-scale implementation with all its organizational and investment consequences should not be undertaken until the pilots deliver clearly superior results. The successful project team will then play a crucial role in training the line managers and in controlling the future planning process.

Depending on the scope of the required process changes and the willingness to make changes and take risks, the time required to implement micromarket management can range from six months to three years.

Serving micromarkets on the basis of customer-oriented analysis of the company's product line has become a key component of the growth and thus the survival strategy of many German companies. The concept of micromarket management offers vital support. It provides the prerequisites to satisfy increasingly sophisticated customer needs and to calculate expected costs and profits realistically. Thus, it supports both top-line revenue growth and bottom-line profit improvement.

References

Ward's Communications Incorporated (1986). "1985 Ward's Automotive Yearbook." Southfield, Michigan.

Ward's Communications Incorporated (1997). "1996 Ward's Automotive Yearbook." Southfield, Michigan.

McKinsey & Company, Inc., G. Rommel et al. (1995). "Simplicity Wins, How Germany's Mid-Sized Industrial Companies Succeed." Harvard Business School Press, Boston, Massachusetts.

G. D. Eppen, W. A. Hanson, and R. K. Martin (1991). "Bundling - New Products, New Markets, Low Risks." Sloan Management Review, Summer 1991, 7-14.

Management Accounting and Product Variety

Volker Lingnau[1]

[1] **Volker Lingnau**, Business School at the University of Mannheim, Mannheim, Germany.

1 Variant Types

If we take a closer look at the phenomenon of variants, it soon becomes apparent that this term merely refers to a generic concept, under which a large number of variant types are subsumed. Different types of variant also cause potentially different costs, so that these variants must first of all be broken down into groups.

A general definition of variants might be "objects normally having a large number of identical components, which exhibit similarities with regard to at least one of the following three attributes: geometry, material and technology". This definition embraces objects with either one or several parts (individual parts/groups), irrespective of whether or not these groups constitute a unit from the planning point of view and of whether or not the objects are destined to be sold on the market (products/components). The possible variant levels can be defined either by the manufacturer or by the customer (manufacturer/customer-specific variants).

Depending on the frequency with which the variants occur, they can be subdivided into 1) variants which occur frequently and more or less regularly, 2) variants whose occurrence is linked either to a trend or to cyclic (usually seasonal) fluctuations and 3) variants which occur only rarely and with a random distribution. Whereas the first and last groups are often described in such terms as "common" and "occasional", the second group is not normally considered separately. To use the terminology of material requirements planning, the variants belonging to the first group are referred to below as X-variants, those in the second group as Y-variants and those in the third group as Z-variants. X-variants in this connection are generally manufacturer-specific variants, while customer-specific variants occur as Z-variants.

If variants also have other variants at the lower structure levels (variants in variants), they are known as *complex* variants; if not, they are called *simple* variants. Single-part variants and multi-part variants which merely differ with regard to one planning unit are therefore bound to be simple variants. If both simple and complex variant structures exist, the variant is referred to as a mixed complex variant (Figure 1).

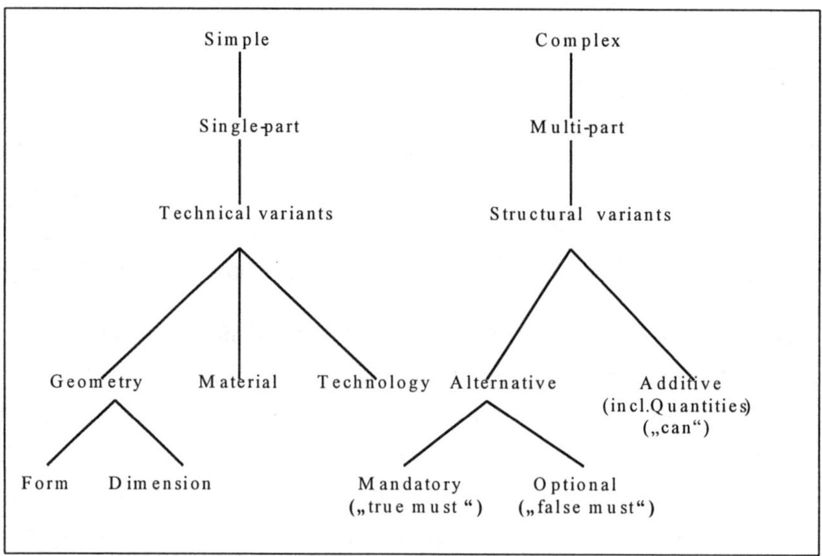

Fig. 1. Structure and attributes of variants

The group of technical variants consists of *geometry, material and technology variants. Geometry variants* are in turn subdivided into *form and dimension variants. Form variants* differ in terms of their external shape and design (e.g. Phillips screws or slotted screws and spherical or cylindrical thermos flasks). *Dimension variants* have the same basic form, but vary with regard to at least one dimension (e.g. M6 x 30 screws or M6 x 40 screws and spherical thermos flasks with a capacity of 0.7 litres or 1 litre). Different tolerance classes could be seen as a special type of dimension variant. *Material variants* are geometrically identical components made of different materials (e.g. screwdrivers made of tool steel or chrome vanadium and cupboards made of solid pine or chipboard with a synthetic surface). One example of a special type of material variant might be *surface variants*, which differ in terms of their colour and structure (e.g. smoothed or polished). Finally, *technology variants* are characterised by the fact that different processing methods and steps are employed at the production stage (e.g. welding or bonding). It is thus clear that technological variants can occur in conjunction with both single-part and multi-part variants, whereby different combinations of the above-mentioned attribute levels are also possible.

If variants can be distinguished in terms of the different components assigned to them, the term structural variant is used. Structural variants can only occur jointly with multi-part variants, as it is not possible for different components to be assigned to single-part variants.

The structural variants are usually classified according to whether a decision *must* be made between various *alternative* components, e.g. between different bodywork colours ("must" variants), or whether the components *can* be chosen additionally, e.g. headrests for the back seats ("can" or *additive* variants). However, this subdivision into must and can variants does not automatically cover situations in which an *alternative* decision *can* be made, i.e. where one component optionally replaces another component (e.g. electric windows instead of manual windows). This situation can occur whenever certain variant levels are offered as *standard.* Depending on the variant's definition, a decision in favour of a particular level can therefore be either *mandatory* or *optional.*

In order to allow a clear distinction to be made, we therefore differentiate between *mandatory alternative variants* ("true" must variants), *optional alternative variants* ("false" must variants) and *additive variants* (can variants). Quantity variants, where the same components are assigned in different quantities, are considered to be a special type of additive variant (see Figure 1).

In order to clarify the relationships that exist between the different variant levels when two variants are compared, we shall refer to a component which is added onto a variant as an *additional component* and to one which is removed from a variant as a *lapsed component.* Strictly speaking, however, mandatory alternative variants do not have additional or lapsed components, because either attribute level A or attribute level B is chosen and not level A instead of level B.

It should also be emphasised that the different variant attributes are often dependent on one another. A distinction can be made between technical and logical dependences. In each instance, both positive dependences (if attribute A is present, attribute B must also be present) and negative dependences (if attribute A is present, attribute B must not be present) can occur, as well as both unilateral and mutual dependences. In the case of technical dependences, the levels of one or more attributes can be used to determine the possible values of another attribute on the basis of physical-technical laws, government regulations or standards. Logical dependences are not the result of the above-mentioned laws, but rather of conventions that are usually company-internal and which (at least theoretically) can be modified at any time.

2 Characteristics and Problems of Diverse Production

Some of the typical characteristics and problems of variant manufacturers in German industry have already been determined within the framework of an empirical study (Lingnau 1994). The most crucial findings were firstly that all those interviewed stated that they produced variants and secondly that as many as 95% would introduce new variants at short notice in the event of changes in demand. These findings were valid not only for the series-production mode of manufacture, but also for one-off, large-series and mass production. Surprisingly, awareness of the problems associated with diverse production declines as the number of variants increases, particularly among small and medium-sized businesses. A rise in the number of variants also leads to a significant growth in production to stock, for example, entailing a risk that the products will become worthless if one of these frequent changes in demand occurs. These findings clearly show the importance of accurate variant costing.

The variant-related problems that were identified in the production area were as follows (in order of importance): the large amount of capital tied up in current assets, the reduction in batch sizes, the increasing number of special and rush orders and the frequent jumps in sales curves.

3 The Effects of Production Diversity on Costs

The effects of production with a large number of variants on costs in the various functional areas are analysed in detail below. The effects on costs in quality assurance and in the warehouse area are analysed separately from this functional classification. The costs in general administrative areas are also affected, for example as a result of the need to take on additional staff, whose records must be maintained in the personnel department, additional purchases of hardware and software, which must be maintained in the computer department, and business valuations of the variants (cost estimates, value analyses).

It is practically impossible to quantify the effects on costs of increasing the number of variants, however, as the costing method does not presently provide the necessary support. A few studies nevertheless reveal that these variant-related costs may well be substantial.

3.1 The Experience Curve for Variant Diversity

According to Wildemann, an inverse experience-curve law applies if the number of variants is increased. As shown in Figure 2, doubling the number of variants causes the unit costs to rise by between 20 and 30% (Wildemann 1990). It would appear useful to examine the effects combined together in the experience curve separately in order to learn more about this phenomenon. A distinction must be made between the fixed-cost degressions per unit (scale effects) on the one hand and the realisation of learning potentials (learning effects) on the other hand (Kloock / Sabel 1993).

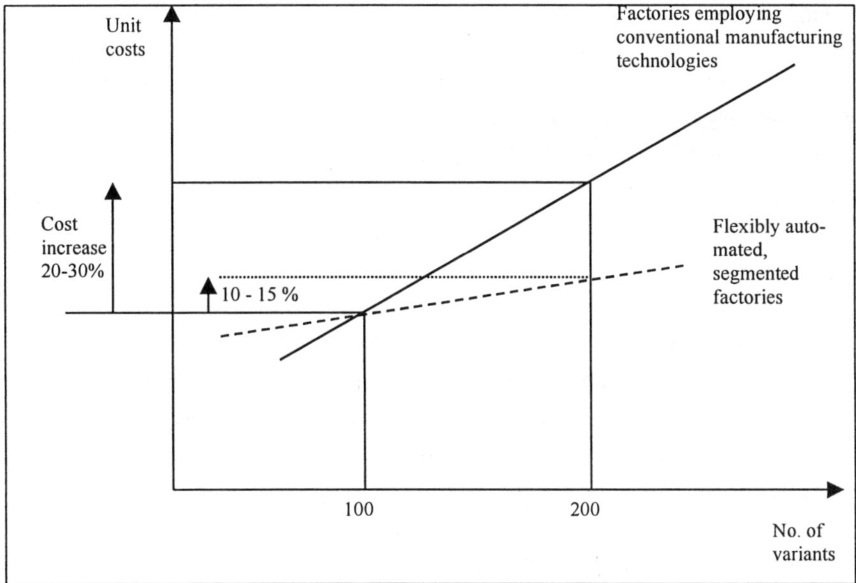

Fig. 2. Unit costs versus number of variants

The scale effects, which are based on a static point of view, can be further subdivided into economies of activity, economies of scale and economies of scope, in other words static unit cost reductions are linked to the utilisation of production capacity, the size of the capacities and the scope of the product range, as well as the process structure. It should be noted that these unit-specific input reductions only occur if there is a parallel increase in output as a result of the economies. The different types of degression effect can be attributed to different assumptions about capacities.

The "economies of activity" degression is based on the assumption that there is exactly one initial capacity and - in its simplest form - that the company is a single-stage, single-product firm. In this case, the variable which causes the unit-cost degression is the capacity utilisation, in other words the period quantity, which in turn is controlled by the marketing mix (and specifically by the selling price).

"Economies of scale" assume that there are alternative initial capacities for a product and a process. The controlling variable is thus the hardware investment, whereby higher investments create higher capacities, leading to lower unit costs in the event of full capacity operation. However, this also means that economies of scale can only be achieved through investments and only if economies of activity are also realised.

With "economies of scope", the assumption is that there are alternative initial capacities for alternative programs and processes, which are controlled by programme and production structure decisions. Economies of scope thus entail a reduction in complexity, whereas an increase in the complexity of more subtly differentiated production ranges is accompanied by the loss of diseconomies of scope.

Learning effects can be categorised in a similar way if they are interpreted as learning capacity, in other words as the realisation of learning potentials over a period of time. It is thus clear that these effects are founded on a dynamic viewpoint. They result both from individual learning processes and from the experience gathered by organisations.

If the effects described here are combined in the experience curve, the following fundamental propositions can be put forward regarding cost reductions through capacity utilisation: existing capacities must be utilised, whereby a comparison of the single and multi-period views is essential. Hardware and software investments are necessary in order to design alternative capacities. Reducing complexity and modifying and designing learning capacity are further important aspects in a multi-product, multi-process company.

The question as to which of these effects predominates can be answered by stating that neither of the two is inherently more significant. It was, however, possible to establish that a relatively small increase in the learning rate means a predominance of hardware investments over software investments, whereas the opposite is true if the learning rate is increased substantially.

3.2 Quantity Effects

An analysis of the cost effects must also consider how introducing a new variant affects the sales figures for the other products, since this to a large extent determines the existence of scale effects and thus the variant-related costs. On the one hand, the introduction of a new variant may be entirely to the detriment of the sales figures for the other products, in other words the total sales quantity remains the same (cannibalisation effect). The only consequence in the areas affected by the production of the variant is a reduction in the quantity produced, while in the unaffected areas there is no change at all. On the other hand, it is also possible that introducing a new variant will have no effect on the sales figures for the other products, in other words the total sales quantity is increased by the quantity of the new variant (bandwaggon effect). No change occurs in the areas affected by the new variant. However, one could also estimate the magnitude of the cost savings that are lost as a result of dispensing with a standardised product. In this case, though, account would at the same time have to be taken of the quantity component, since the increase in the sales volume is a result of precisely this differentiation. In the areas that are common to both the new variant and other variants, the larger quantity leads to positive scale effects.

Both effects can normally be observed in practice, as although the total quantity increases, the rise is less than the quantity of the new variant. Empirical studies have revealed a predominance of the cannibalisation effect. For this reason, and in order to simplify our explanations, we shall assume a pure cannibalisation effect in the following.

The aspects described here not only have an effect on costs, of course, but also on revenues. At first glance, it would seem a simple matter to assign these; they can be derived from the price list (at least in the case of manufacturer-specific variants). However, it is also necessary to consider the quantity component (see above) in addition to the unit prices, particularly if the product has a large number of variants, since diverse dependences may exist between these variants. In addition, sales volume forecasts are extremely difficult when a new variant is introduced, as the distribution of the quantities sold between the individual variants is highly uncertain, even though the projected total quantities may well be correct.

Moreover, the experience acquired when a variant is introduced cannot necessarily be applied when a variable is withdrawn simply by inverting the sign. A buyer who used to purchase another product in the same range before the new variant was introduced may not return to the original product after the old variant has been withdrawn, but may instead choose a similar product from another manufacturer.

3.3 Cost Effects in the Purchasing Area

3.3.1 Purchasing Costs

An increase in the number of variants is always accompanied by an increase in the number of different components that must be purchased, irrespective of the concrete attributes of this new variant, so that purchasing activities become more complex. If alternative variants are offered, there is also a drop in the quantity that must be purchased of each of the components concerned.

Purchase costs are normally subdivided into quantity-dependent costs (variable, direct) and quantity-independent costs (fixed-order, indirect). Since, however, as explained below, some of the costs that are usually considered to be quantity-independent are - at least in part - in fact also quantity-dependent, we shall make the following distinction between direct and indirect costs: direct costs are those costs which are obtained by multiplying the quantity procured by the acquisition cost per unit. All direct costs are thus quantity-dependent. Indirect costs are those costs which are incurred within the purchasing company as a result of processing the order. These costs are predominantly fixed-order costs, though they may also include quantity-dependent elements.

1) Direct Costs

Smaller quantities per component often mean less favourable terms and conditions, and in particular the loss of bulk discounts or the acceptance of extra charges for small amounts. This applies not only to the market prices of the purchased components, but in some cases also to the costs of transport, packing and insurance. Quantity-dependent cost graduations of this kind result in a disproportionate - or even abrupt - rise in costs despite falling quantities. An increase in direct (unit) costs is therefore only to be expected if alternative variants are offered. Additive variants, on the other hand, cause the total required quantity to rise, so that the effects on direct costs may even be positive.

2) Indirect Costs

The increase in the number of ordered items necessitates additional effort for coordinating and monitoring orders, entailing a greater likelihood that part deliveries will have to be accepted because the supplier is not able to deliver some of the components immediately. Each new component means negotiations with existing suppliers and possibly a need to select new suppliers. As the number of different components increases, so does the risk of material planning errors, leading to shortfalls with in some cases far-reaching consequences for costs. This was also confirmed by the empirical study. Similarly, if too many components are ordered, additional costs are incurred as a result of inflated stocks. This risk is further aggravated by the deterioration in forecast quality as the number of variants is increased.

An increase in the number of variants also results in an increase in the amount of work involved in the receiving inspections. Deadline and quantity controls are rendered more complex as a result of the greater variety of components, and the problem is added to by part deliveries. The effort necessary for the most important, and most extensive, step in the receiving inspection, namely quality control, is affected both by the increase in the number of different components and by the fall in the quantity per component (see below). If the components are obtained from a new supplier, more detailed controls (e.g. a 100% inspection) will often need to be carried out on the first few deliveries than is the case with existing suppliers whose quality is already known.

If special components are required for a new variant instead of standard components, it may be necessary for these components to be manufactured in-house rather than purchased from outside suppliers like the standard components. The opposite situation is also conceivable, in other words the infrastructure needed for production (e.g. know-how or machines) may not be available within the company, so that the components have to be purchased from outside suppliers. The exact effects on costs of in-house production or purchase from outside suppliers can only be assessed on a case-to-case basis. It is generally true to say, however, that if only some of the work steps are contracted outside, in other words if the coordination services remain the responsibility of the company, the result will be a disproportionate rise in costs. If, on the other hand, complete work packages, including all auxiliary services (such as packing and transport), are contracted to outside companies, a reduction in variant-dependent costs is likely.

3.3.2 Shortage Costs

The increase in the number of variants is accompanied by an increased probability of shortfalls, with the result that it is not possible to meet material requirements, so that shortage costs are incurred. These consequences may affect the costs of purchasing, production and sales. The shortage costs can also be subdivided into direct and indirect costs. Direct shortage costs are those costs that result from the shortfall itself, while indirect costs are the costs that ensue when action is taken to avoid this shortfall.

1) Effects on Purchasing

Direct shortage costs may result for purchasing if additional activities become necessary as a result of the shortfall (e.g. performing new quantity calculations or selecting alternative suppliers). If the shortfall leads to production downtimes, the affected components will remain in stock for longer than planned, so that additional costs are incurred. Further costs may then ensue if the warehouse space which is blocked in this way is also required for other components, for which other warehouse space now has to be found.

Once the company becomes aware of the risk of a shortfall, indirect shortage costs may be incurred as a result of efforts to avoid this situation, such as additional transport costs (express orders), costs of purchasing the component concerned - or a more expensive, alternative component - from another supplier on less favourable terms and conditions.

2) Effects on Production

Direct shortage costs may result for production in the form of downtime and restart costs. Indirect shortage costs may be incurred as a result of efforts to make up for production that was lost as a result of the shortages, for example by adjusting production times and intensity, by putting standby machines into operation or by outsourcing orders at short notice.

If adjustments are made to production times by introducing overtime, there will be a sharp increase in the variable manpower costs due to the bonus payments that must be made for overtime hours. If additional shifts are operated, fixed costs will also rise. If adjustments are made to the production intensity, progressively rising variable costs must be expected. Using standby machines (which are frequently older) may lead both to higher variable unit costs and to higher fixed costs as a result of the quantitative adjustments.

If orders are outsourced at short notice, it is important to remember that there will not only be additional costs for purchases from outside suppliers, but possibly also quality differences, which may lead to further costs for essential re-working.

3) Effects on Sales

Costs may be incurred in this area if delivery deadlines are endangered as a result of the shortfall. The majority of direct shortage costs take the form of reduced contribution margins and contract penalties. In any case, there is likely to be a loss of goodwill, albeit one which is difficult to quantify. Indirect shortage costs may ensue as a result of express deliveries to prevent deadlines from being exceeded as well as the costs associated with repeat orders and separate deliveries.

3.4 Cost Effects in the Production Area

The effects of a large number of variants are usually particularly great in the production area, as manufacturing costs are extremely sensitive to diversity. The additional costs are due mainly to the increasing complexity of the planning and control tasks as well as to the higher number of batch changeovers.

3.4.1 Planning and Control

Coordination costs rise as the processes become more complex, because a larger number of variants necessitates more detailed planning. An extensive control system, capable of ensuring the supply of individual parts, is required to control the flow of materials. Furthermore, an increase in the number of variants automatically means constantly changing bottlenecks, so that machine loading schedules remain valid for a correspondingly shorter period of time. In addition, there are always idle capacity costs in one part of the production plant. The more detailed the planning stage and the greater the dependences between the individual operations, the more difficult control becomes in the event of unforeseen variances. Often only limited adjustments are possible, so that idle times occur. Finally, processing steps necessary for the production of variants are frequently introduced or modified at short notice, without adequate account being taken of their dependences with other operations or of the consequences for other areas within the company. The resulting costs - particularly for quality assurance and re-working - are significantly higher than for general organisational adjustments.

Special problems are created as a result of increasing the number of variants when it comes to balancing assembly lines. The cycle times are fixed by estimating the probable frequency of occurrence of each variant and defining a variant sequence (model mix). If the estimated frequency does not coincide with the daily schedule, either there will be idle times or more than the planned number of staff - usually well-qualified, so-called stand-ins - will be required (static case). The same consequences result from a deviation in the defined sequence (dynamic case). An increase in the number of variants causes assembly times to vary considerably and leads to a rise in the number of static and dynamic variances. Shortfall situations and missed deadlines follow in addition, leading in turn to the build-up of large stocks between production and assembly as well as in the assembly plant itself, accompanied by an increase in the number of defects. The importance of this phenomenon was confirmed by the empirical study.

3.4.2 Batch Changeovers

If we ignore the no doubt rather theoretical possibility that each variant is manufactured using production equipment specifically tailored to it, it is evident that the number of batch changeovers (set-up procedures) increases with the number of variants. These may necessitate only minor interventions in the production process or they may be extremely complex, for example when a new model is introduced in the automotive industry, where preparation and implementation may take several months (Kaluza 1989, p. 41 ff.).

The costs which ensue from a batch changeover can be subdivided as follows according to their causes (influencing variables):

1) Preparatory Operations

With the exception of fixed standby costs, especially in the indirect areas, production preparation costs are incurred as a result of both planning and implementing activities, such as purchasing, making available and processing production documentation (such as parts lists or work schedules), training staff and providing materials and work equipment/production facilities. The latter type of costs also leads to corresponding inventory costs. The larger the number of variants, the more complex planning and implementation become, since the volume of available experience declines with each new variant. The number of identical activities is thus reduced and preparation times are longer. The time required to train staff also increases, as there are more different production steps that have to be mastered. Additional costs are sometimes incurred because better-trained staff is necessary to perform more complex activities. Particularly in the case of customer-specific variants, the need for special measures should be examined, for example on account of oversize, as the normal production process is not suitable for certain attribute levels.

2) Phase-out Operations

When a batch changeover takes place, the production facilities must normally be shut down either partly or completely, meaning that the previous production process and the activities it involves (e.g. lowering tools, releasing pallets) is discontinued. The phase-out operations required for this purpose may be extremely complex, especially if they affect interlinked production facilities and necessitate additional staff and energy, for example for braking operations or pressure stabilisation. Furthermore, it is likely that the wear and tear on machines and tools will be greater than during normal operation and that more scrap will be produced. The workpieces and tools must then be removed and transported to their destinations.

3) Change-over Operations

The costs of change-over operations (direct set-up costs) are the costs which arise directly from preparing a production facility for the fulfilment of a particular task (setting up). In the case of more complex production facilities, especially conventional transfer lines, change-over operations in the form of modification activities may necessitate considerable time and staff resources. Such activities must usually be carried out by highly qualified workers (machine set-up specialists). The additional consequences of repeated set-up procedures caused by the large number of variants means that the effort for setting up the machines is greater, because no learning effects are possible. Furthermore, measures may have to be taken to safeguard work safety.

4) Cleaning Operations

Cleaning work on production facilities or workpieces is often necessary after a batch change-over, for example in order to remove paint residues. This work results in costs for wages, materials and energy.

5) Start-up Operations

Start-up operations take place during the period from switching on the production facilities to reaching the normal level of performance. In the same way as the phasing-out operations, these operations are likely to necessitate a higher energy consumption for reaching the process conditions (e.g. speed, temperature, pressure), as well as additional wear and tear and a higher scrap rate. Once again, special problems can occur if the facilities are interlinked. Furthermore, the operators require a period of familiarisation after each batch change-over, during which the defect rate will be higher. More extensive monitoring than usual is essential in order to avoid failures.

6) Production Disruptions

Unless the production facilities allow setting up while the machine is still running, batch change-overs cause disruptions to production in the form either of a complete shut-down or at the very least restricted equipment availability. All the factor inputs that are necessary to maintain readiness for operation must continue to be made throughout the production disruption, which lasts from the start of the run-down phase to the end of the start-up phase. Any factor costs that arise because other production facilities operate less efficiently as a result of the batch change-over must also be assigned to this change-over.

Finally, contribution margins may be reduced owing to the downtimes of the production facilities during the setting-up procedure, if an alternative use might otherwise have been possible. These reduced margins take the form of opportunity costs, in other words unlike the other types of cost effect discussed so far they are not characterised by factor inputs when the machines are set up.

3.5 Cost Effects in the Sales Area

3.5.1 Innovation Phase

1) Idea Phase

During the idea phase, ever more sophisticated market research models must on the one hand be used to determine new wants on largely saturated markets. As a result, specialists are required to prepare, carry out and evaluate the studies, and a considerable burden is placed on the respondents, so that correspondingly high costs are incurred for remunerating expenses. On the other hand, identifying new wants is also becoming increasingly difficult - whether this process entails determining as yet vacant market niches on the basis of market research or creating completely new wants. Selecting ideas is becoming more and more complicated on account of the large number of dependences between the individual variant levels, especially as the use of systematic evaluation schemes is rendered additionally problematic by what are often only minor variations. These aspects have no major consequences as far as costs are concerned, however, as the idea phase only contributes a very small share towards the total costs for product planning.

2) Concept Phase

The concept phase entails costs for drawing up the system specification and for analysing the prospects for success. The latter task is becoming increasingly complex owing to the large number of dependences between the growing number of different variants.

3) Development Phase

The variant development phase may give rise to costs that differ widely, depending on its scope (e.g. development of chrome-plated door handles as an optional extra versus an electronic anti-blocking system). For this reason, it is only possible to describe the potential cost effects of an increase in the number of variants in very general terms here.

More variants means additional work when new elements are designed, as well as in connection with the production and administration of the necessary technical documentation. The newly developed parts result in extra costs in relation not only to their design, but also to the essential, regular maintenance, such as adaptations to take account of technical advances. These costs differ according to the level on which the changes are made and may well be substantial, especially for complex variants. The rise in the number of different parts entails a corresponding need for coordination, as it is important to ensure that no undesirable dependences are created between new and existing variant levels (the motor for the new electric window might itself fit easily into the existing door, for example, but not together with the heat-absorbing panes or the loudspeakers). Furthermore, even minor alterations can sometimes necessitate extensive acceptance procedures that are prescribed by law. This problem is intensified by the differences between national regulations and requirements. Additional costs arise if existing production methods and tools need to be adapted or new ones developed, as well as for enhancing the performance of the computer system, for example if CAD needs to be used.

Very little is known about the development costs that arise in connection with variant diversity, however. Moreover, they are allocated in a manner that is not only inadequate but also distorts the results, especially since traditional cost accounting does not normally take effect until after the end of the design phase, so that standard products are burdened with costs that are too high, while subsequently developed variants are costed too low.

4) Test Phase

Testing the developed ideas represents the largest pool of costs within the product planning framework. The prototypes are tested first of all, in order to permit any design defects to be identified and remedied; a pilot lot is used to verify whether or not the product can be manufactured to the required quality with the available production facilities and methods. A market test is then conducted to establish whether the product that has been developed is also sellable and which instruments should be employed to back up sales.

The variant-related cost effects of these extremely diverse tests are due in part to the fact that it is not normally sufficient simply to perform type-specific tests, but on the contrary each variant must be tested separately, albeit to a differing degree. A large number of variants means, however, not only that a correspondingly large number of tests have to be conducted, but also that considerable effort is necessary to verify the large number of potential (variant-related) dependences. Even if it is possible to make use of empirical data when designing adaptations and variants, the initial acquisition of this data and keeping it up to date entail costs.

3.5.2 Market Life Cycle

The decisions concerning the marketing mix that must be taken at the time of the market launch become increasingly complex the larger the number of variants. Those responsible for pricing policies are obliged to estimate the costs of an ever greater number of products, and to take account of both their own and rival products when positioning them. The communications policy must be suitable for making the individual variants, and particularly the differences between them, known to the target segment. In connection with the distribution policy, it may be necessary to design the sales channels and the distribution logistics on a variant-specific basis. New dealers may have to be secured - something which is often only possible by offering promises of extensive sales support, special terms and conditions or even payment of listing fees. Repurchase guarantees in the event that a particular sales threshold is not reached are also possible. Moreover, it is unlikely that additional shelf space, previously occupied by rival products, will be allotted to the new product variant, in other words the shelf space for the manufacturer's other products will be reduced. These aspects are further exacerbated by the fact that up to 95% of all new product launches fail.

If the product range is distributed via the company's own field service, costs will initially be incurred for staff training and sales documentation. The success of the field staff's activities is limited by the number and variety of the products it is required to represent. The greater the number of variants for which a sales representative is responsible, the longer he will spend visiting each dealer, which means that either the staff level must be increased or the frequency of the visits reduced, with a corresponding deterioration in their effectiveness. However, in this connection we should also mention the risk that the field staff will put together their own personal product ranges, which they think will be either easiest to sell or easiest to present. Along with the limited storage space of the dealer, the above-mentioned reduction in the shelf space available per product means more frequent orders consisting of more items, but at the same time smaller quantities for each item, so that the amount of work necessary to process each order rises accordingly. At the same time, the risk of part deliveries increases. There is a tendency among sales staff to favour high closing inventories, as a means of preventing supply shortages.

If the company maintains its own customer service department, it too will be faced with rising costs due to the higher number of spare parts and variant-specific tools that need to be stocked. Warehousing costs do not arise merely during the variant's life cycle, but for as long as this variant continues to be supported by customer service. This period may well be considerably longer than the market cycle. A figure of at least ten years after a model has been phased out is quoted for car manufacturers. Training for customer service staff and provision of up-to-date customer service information are further cost factors, because the scope of customer service literature - and thus the effort necessary to process and administer it - increases together with the number of variants.

3.6 Cross-functional Cost Effects

3.6.1 Quality Assurance Costs

The costs for quality assurance in the sales, production and purchasing areas rise if the number of variants is increased. Quality standards must be defined for new components jointly with the design departments prior to the start of production, on the basis of the targets determined by means of market research. These standards must be agreed upon with the suppliers and a suitable monitoring system set up. If a 100% inspection is carried out, the reduction in the quantity per component means a corresponding reduction in component-specific inspections. Apart from possible effects on the experience curve, it therefore has no influence on the overall inspection outlay. If only a random sample is inspected on the other hand, there is no proportionality between the quantity per component (size of the sampled population) and the size of the sample (Geiger 1986, p. 179). In the majority of cases, the sample size is in fact independent of the size of the sampled population. A reduction in the quantity per component would thus mean no change in the volume of inspection outlay for each individual component, but the increase in the number of different components would mean a rise in the overall inspection outlay.

Let us consider the production of - to date - 8000 units of component A 1 per month as an example. A random sample of these units is to be inspected on the basis of qualitative characteristics (inspection by attribute). Despite its weaknesses, this inspection method based on qualitative characteristic sampling has gained acceptance in practice over inspection procedures based on quantitative characteristics (inspection by variable). It has become even more important for receiving and final inspections than for production, being regularly included in supply agreements. A minimum sample size of 367 has been determined for the inspection. As a result of introducing an alternative variant, only 4000 units of component A 1 are now required, along with 4000 units of component A 2 for the new variant. This means that a sample of 367 units must in future be taken for the inspection of each of the two deliveries. Although the total quantity has not changed, the inspection outlay has thus been doubled. If the quantity per component is reduced too far, it will be necessary to increase this outlay even more by changing over to a more complicated inspection procedure. This will also mean a need for added controls on account of the greater risk of defects, as explained earlier.

3.6.2 Inventory Costs

A basic distinction can be drawn between incoming, intermediate and outgoing stores. These are usually allocated to the purchasing, production and sales areas, whereby similar types of problem are encountered with each of the three different stores.

1) Value-dependent Inventory Costs

An increase in the number of variants generally also means an increase in the total inventory even in conjunction with a pure cannibalisation effect. This is due firstly to the level of the reserve inventory. This level is dependent above all on the accuracy of the demand forecast and the reliability of the supplier. The accuracy of the demand forecast declines as the number of variants increases, so that the necessary reserve inventory level rises. Furthermore, an increase in the number of variants also entails a risk of deliveries arriving too late or in insufficient amounts, both of which cause the reserve inventory level to rise. Secondly, a reduction in the demand per component merely leads to a disproportionate fall in the order quantity and thus in the average inventory, so that if the existing quantity is now subdivided into two components, there will be an overall rise in the average inventory level.

If alternative variants are introduced, the effects described above result in a relative increase in inventories of existing components. An increase in the total inventory (or in its value) can also be brought about if the new variant level contains more components (or components with a higher value) than the previous level. If additive variants are introduced, there is no reduction in the required quantities of the old components, so that the above-mentioned effects do not apply. There will nevertheless always be an increase in the total inventory on account of the additional component.

The increase in the inventory levels and the periods of storage is accompanied by an increased risk that the value of the inventory will be impaired as a result of price reductions, shrinkage or acts of God. There is also a risk that the inventory will become obsolete due to technical advances, especially if components are stocked for Z-variants for which there is very little demand. This problem was made quite plain by the empirical study.

2) Quantity-dependent Inventory Costs

The inventory-increasing effects described above of course also influence quantity-dependent inventory costs. The effects on costs of occupying warehouse space (depreciation, interest, insurance, rent) and of maintaining the warehouse (lighting, heating, ventilation), in relation to the activity level although these are generally considered to be fixed (activity) costs, must however also be taken into account.

A separate storage bin with an appropriate space requirement is normally needed for each component that must be stocked. This space requirement is usually only available in certain sizes (e.g. standardised skeleton boxes), so that a fixed space requirement rising in steps ensues. An increase in the number of different components can thus result in a significant rise in the required storage space, making extensions necessary, even if the quantity that needs to be stored remains the same.

3) Movement-dependent Inventory Costs

An increase in the number of storage bins also leads to an increase in the times for storage and disbursement, since the distances that must be covered are longer and the process of finding a suitable bin is slower, although this is likely to be less of a problem with computer-aided warehouse management systems.

4) Material Planning Costs

An increase in the number of different components also means a corresponding increase in the number of articles in stock that must be scheduled. As a result, inventory management becomes more complex and material planning costs rise.

4 Variants and Cost Determinants

It can be concluded from our above observations that the normal situation, where a large number of variants are produced, leads to cost increases in practically all areas of a company. There are cost effects notably in connection with administration, preparation and development. Both planning and implementation are affected in the functional areas. As a result, shifts can be observed in the cost structure, leading to an increasingly large proportion of indirect costs on account of the additional work which is necessary for planning, control, monitoring and coordination as opposed to the actual production process (Hoitsch 1997, p. 55). This finding underlines the great importance that needs to be attached to cost effects when decisions are taken regarding the production programme, even if the changes in question are (apparently) only minor.

Modern methods of marginal costing consider the activity level as dominant cost determinant, and attempt above all to subdivide costs according to their quantitative yield. This output-oriented subdivision into variable activity costs and activity costs has a deep stratification in the direct areas. Planning and control are supported with decision-oriented cost information by means of a mature system of allocation bases. Readiness for operation is normally also taken into account as a cost determinant. Costs for maintaining this readiness are incurred as a result of the available capacities. They thus constitute an input-oriented cost determinant.

In the indirect areas, on the other hand, volume oriented activity can only be taken as a cost determinant for a small portion of the costs (Miller / Vollmann 1985). These costs are primarily determined not by the output, but by the throughput, in other words by the processes that take place in the area in question. The system of cost determinants must therefore be extended to include the complex aspect of "work contents". The work contents determinant must be operationalised with the aid of suitable allocation bases in the same way as the volume oriented activity determinant. These variables might be the number of processes if the work contents are homogeneous (e.g. the number of orders) or the process time for heterogeneous work contents (e.g. the time required to process an order). The number of variants could also be taken as an allocation base, though of course the variants must be comparable if they are to serve as standard bases. The variants would in this case have to be subdivided analogously to the system described earlier.

The above observations concerning cost determinants are summarised once again systematically in Figure 3. They permit the following assertions to be made regarding the relevance of the identified cost determinants for operational decisions:

- Activity (output-oriented cost determinant): The variable activity factor costs are relevant for quantity-related production programme decisions, providing both the composition of the programme and the capacities remain constant (quantitative programme decisions).

- Work contents (throughput-oriented cost determinant): The variable process factor costs occur (largely) independently of the output, and are thus irrelevant for quantitative programme decisions. They are relevant, on the other hand, for decisions concerning the composition of the production programme (qualitative programme decisions, e.g. introduction of new variants).

- Capacities (input-oriented cost determinant): The variable capacity factor costs occur independently of both the output and the throughput, and are thus irrelevant for short-term decisions where constant capacities can be assumed.

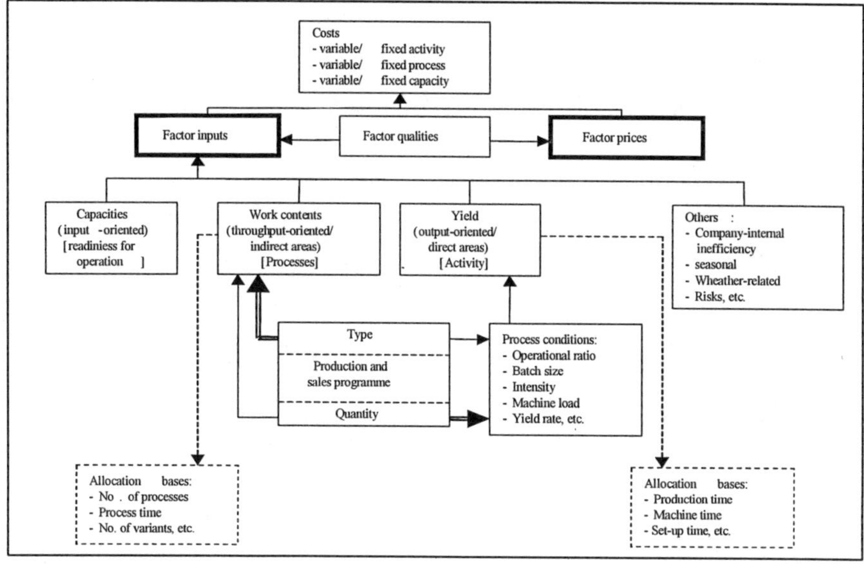

Fig. 3. System of cost determinants

References

Cooper, R and R. S. Kaplan (1988). "Measure Costs Right: Make the Right Decisions." Harvard Business Review, (Sept. / Oct), 96–103.

Geiger, W. (1986). Qualitätslehre. Braunschweig, Wiesbaden.

Hoitsch, H.-J. (1997). Kosten- und Erlösrechnung. 2nd edition, Berlin etc.

Kaluza, B. (1989). Erzeugniswechsel als unternehmenspolitische Aufgabe. Berlin.

Kloock, J. and H. Sabel (1993). "Economies und Savings als grundlegende Konzepte der Erfahrung." ZfB 63, Vol. 3, 209 – 233.

Lingnau, V. (1996). Variantenmanagement. Berlin.

Miller, J. G. and T. E. Vollmann (1985). "The Hidden Factory." Harvard Business Review, (Sept. / Oct.), 142–150.

Wildemann, H. (1990). "Die Fabrik als Labor." ZfB 60, Vol. 7, 611 – 630.

A Conjoint Analysis-Based Procedure to Measure Reservation Price and to Optimally Price Product Bundles

Georg Wuebker[1], and Vijay Mahajan[2]

[1] **Georg Wuebker**, SIMON, KUCHER & PARTNERS, Bonn, Germany.
[2] **Vijay Mahajan**, Department of Marketing, The University of Texas, Austin, Texas.

1 Introduction

Many firms market multiple product lines. Each firm is faced with the problem of deciding whether it should offer products separately (pure component strategy) or whether combinations of products should be marketed in the form of "bundles" or "packages" at a "bundle price" (pure bundling strategy). The use of mixed bundles where buyers have the choice to buy either the bundle or the individual items has recently become increasingly popular. This is the so-called mixed bundling strategy (Guiltinan, 1987). The numerous examples of product bundles include:

- Fast food restaurants, who offer value meals at a special discount.
- Software companies, who offer software packages at a special price.
- Airlines, who bundle vacation packages by combining airfare with lodging.
- Automakers, whose cars are offered in combination with different options such as air conditioning, power windows, leather seats, or central locking.

In all of the above examples, companies are interested in knowing how to price the bundled products or services. Specifically, they are interested in questions such as:

- How do we determine the optimal prices of the bundled products?
- How sensitive are profits to variations in the bundle prices?

As reviewed in the next section, several suggested approaches to answering the above questions are based on the estimation of the reservation price or the maximum price that a consumer is willing to pay for a specific bundle type. These approaches typically ask consumers to indicate their reservation prices directly (Hanson and Martin, 1990; Venkatesh and Mahajan, 1993). This direct method has several disadvantages (Simon, 1989, 27; Monroe, 1990, 107).

First, price-related questions may induce an unrealistically high price-consciousness among respondents. Second, the price is evaluated in isolation, whereas in reality, consumers weigh the price against other attributes and make a trade-off. Another disadvantage of the direct method is that buyers may attempt to quote artificially lower reservation prices, since many of them feel obligated to act as conscientious buyers who help to keep prices down (Morton, 1989). Therefore, the direct method may yield incorrect reservation prices, which lead to the incorrect estimations of overall demand and price elasticities. By relying on these misleading reservation prices, companies will estimate sub-optimal prices and hence lose profit.

Conjoint analysis - a decompositional procedure which measures consumers' trade-offs among multi-attribute products and services (Green and Srinivasan, 1990) - can be used to avoid these shortcomings. Despite its importance, we know of only one article in the marketing literature (Kohli and Mahajan, 1991) that shows how to estimate reservation prices via conjoint analysis. However, their approach is limited to the estimation of reservation prices for individual items.

The objective of this paper is to propose a conjoint analysis-based procedure to measure reservation prices for different items and bundle types. We empirically compare the conjoint analysis-based procedure with the direct method. From this comparison, we establish the superiority of the conjoint analysis-based approach over the direct approach in predicting respondent preferences for various bundle types; obtain better estimated reservation prices for bundles of products and; hence, realize higher profits.

In the following section, we briefly review the main literature on price bundling. In the third section, we present the model development. We then show how to estimate reservation prices for a bundle type in the fast food industry. After presenting the results, we conclude with a discussion of limitations and directions for future research.

2 Literature Review

Previous research in bundling primarily provides theoretical rationales for bundling. An overview of this literature is provided by Simon and Wuebker (1998, see Table 3). Bundling issues in marketing can be divided into two streams of research. First, since the early 1990s, many researchers have applied several theories (prospect theory, mental accounting, information integration theory) to understand consumers' evaluations of bundles (e.g. Gaeth et al., 1990; Drumwright, 1992; Yadav and Monroe, 1993; Yadav, 1994; Kaicker, Bearden and Manning, 1995). For example, Yadav and Monroe (1993) examined how buyers' perceived transaction value of the bundle is influenced by the magnitude of savings offered on bundles and on individual items. Their findings indicate that additional savings offered directly on the bundle have a greater relative impact on buyers' perceptions of transaction value than savings offered on the bundle's individual items.

A second research stream has focused on extending the economic analyses of the sellers' rationales for bundling. Recently, some researchers have focused on optimizing the design and pricing of bundles (e.g., Hanson and Martin, 1990; Venkatesh and Mahajan, 1993; Fuerderer, Huchzermeier and Schrage, 1994). For example, Hanson and Martin (1990) developed a mixed integer programming approach which is useful in identifying the optimal bundle size and price. However, reservation prices of consumers for all possible bundles, the size of the customer segments, and the cost of supplying customers in a specific segment with a certain bundle must be known to implement their approach. Venkatesh and Mahajan (1993) proposed a probabilistic approach that enables sellers to determine optimal prices of a bundle and/or its component products under pure components, pure bundling, and mixed bundling strategies. They considered a season ticket bundle for a series of entertainment performances such as sports events and music concerts. They assumed that consumer purchase decisions were a function of two independent dimensions: time to attend performances and reservation price per performance. They further assumed that the individual reservation prices follow a specific probabilistic distribution which is related to the demand function, which in turn is used for the profit function. Based on this information, they estimated the optimal prices of the bundle and components (individual performances), and the corresponding maximum levels of profits under each strategy.

However, in the approaches suggested by Hanson and Martin (1990) and Venkatesh and Mahajan (1993) respondents were asked to indicate their reservation prices directly. This direct method has several shortcomings, as discussed in Section 1. Therefore, we suggest a conjoint analysis-based approach to measure reservation prices.

3 The Proposed Approach

3.1 Conceptual Underpinnings and Problem Setting

To explain the conceptual underpinnings of the proposed approach, let us consider consumer preferences for different fast food items or bundles. Let us assume that the fast food restaurant McDonald's offers French Fries, a Big Mac and a Drink as individual items. Let us further assume that the restaurant is interested in offering a bundle of these three products. An important question is: if the restaurant has already fixed the prices of individual items, what should it charge for the bundle?

To answer this question we consider the bundle/item type and the price as the two relevant attributes in a conjoint analysis problem setting. For example, given three individual items such as French Fries, a Big Mac and Drink, it is possible to generate seven (= 2^3 - 1) different levels of bundle/item type: French Fries; Big Mac; Drink; French Fries and Big Mac; French Fries and Drink; Big Mac and Drink; and a bundle of all three items.

Table 1. Relevant attributes and levels

Bundle/Item Type (levels)[a]	**Price (levels) in Deutsche Mark (DM)**
French Fries (large) (FF)	2.50
Big Mac (BM)	5.00
Drink (medium) (D)	7.50
French Fries (large) and Big Mac	9.00
Three-Item Bundle (FF and BM and D)	10.50

[a] For the sake of simplicity, we consider only five instead of seven (= 2^3 - 1) possible items.

For the sake of simplicity, Table 1 lists five levels of bundle/item type. The combinations of French Fries and Drink, and Big Mac and Drink are excluded. Table 1 also lists five different levels of price in Deutsche Mark: 2.50, 5.00, 7.50, 9.00 and 10.50. Although it is theoretically possible to generate 25 combinations (= 5 x 5) from the combinations of the various levels of these two attributes, Table 2 lists seven feasible stimuli which assume that the prices of four levels of bundle/item type are fixed: Drink at DM 2.50, French Fries at DM 2.50, Big Mac at DM 5.00, and French Fries and Big Mac at DM 7.50. For the restaurant to maximize its overall profit, one question now is: Should it offer the bundle of all three items at DM 7.50, DM 9.00 or DM 10.50?

Table 2. Profile description of seven feasible stimuli

	Attributes	
Stimuli	**Bundle/Item Type**	**Price (in DM)**
1	Drink	2.50
2	French Fries	2.50
3	Big Mac	5.00
4	French Fries and Big Mac	7.50
5	French Fries, Big Mac and Drink	7.50
6	French Fries, Big Mac and Drink	9.00
7	French Fries, Big Mac and Drink	10.50

To answer this pricing question, we assume that each buyer has a maximum price that he or she is willing to pay for the three item bundle. We also assume that this reservation price is determined by his or her estimated utility for the three item bundle in relation to the price and utility for his or her most preferred bundle type (items) among all bundle type offerings (in our example the 7 stimuli) in his or her evoked set.

3.2 Analytical Underpinnings

We assume that the conjoint analysis data have been collected for the two attributes (bundle/item type, price). The preference model is specified in the following equation.

$$U_k = \sum_{i=1}^{2} \sum_{j=1}^{5} X_{kij}\beta_{ij} + \varepsilon_k , \qquad (1)$$

where U_k = preference score (total utility) for stimulus k,
X_{kij} = dummy variable with presence (=1) or absence (=0) of level j in attribute i for stimulus k,
β_{ij} = part-worth utility associated with level j of attribute i,
ε_k = error term for stimulus k.

In our example, we consider only seven stimuli (see Table 2). From equation (1) we see that we have to estimate 8 part-worth utilities. However, we cannot use a traditional full-profile approach to estimate these part-worth utilities, since we only have 7 observations. To avoid this problem, we use the ACA (Adaptive Conjoint Analysis; Johnson, 1987) procedure - a hybrid conjoint analysis which is used extensively in commercial applications. Wittink, Vriens and Burhenne (1994) document that 42 percent of commercial applications apply ACA for the stimulus construction. The ACA data collection and part-worth estimation procedure as well as the advantages and disadvantages of this approach are presented and discussed in several articles (Agarwal and Green, 1991; Green, Krieger and Agarwal, 1991; Johnson, 1991). The individual part-worth utilities are estimated by an OLS regression updating procedure. Details of this procedure can be found in Green, Krieger and Agarwal (1991) and Johnson (1991).

To show how ACA can be used to assess reservation prices for the three-item bundle in our above example, define u_m* as consumer m's highest estimated total utility of any currently available stimuli. The value of u_m* includes the part-worth utilities of the two attributes bundle/item type and price. Let B denote the index of the three item bundle for which the reservation price is to be assessed. Let $u_{mB|\sim p}$ denote the part-worth utility of the three item bundle, where the notation emphasizes that the utility contributed by price is not included. We assume that consumer m prefers the three item bundle B at price p to any available stimulus if equation (2) is satisfied.

$$u_{mB|\sim p} + u_m(p) = u_m{}^* + s\,, \qquad (2)$$

where $u_m(p)$ denotes the part-worth utility of price p and s defines an arbitrarily small positive number. Based on this equation, p_{mB} is the estimated reservation price of the three item bundle for consumer m if $p = p_{mB}$ is the price at which equation (2) is satisfied.

Table 3. Estimated part-worth utilities of two attributes for one consumer

Price (levels) (DM)	Utility	Bundle/Item Type (levels)	Utility
2.50	35	French Fries (large)	15
5.00	30	Big Mac	23
7.50	20	Drink (medium)	0
9.00	15	Big Mac and French Fries	35
10.50	0	French Fries, Big Mac and Drink	55

An example should make this estimation process clear. Assume that the estimated individual part-worth utilities obtained via a conjoint analysis-based procedure are those shown in Table 3. Note that the utility for prices in the table are positive because they are scaled to be positive numbers. This does not affect the analysis since the utilities in a conjoint analysis are unique only to a linear transformation. From Table 3 we can create the overall utility for the seven stimuli listed in Table 2. These overall utilities are reported in Table 4. The highest overall utility for this consumer is given by the three item bundle at a price of DM 7.50. The overall utility (75) is obtained by adding the part-worth utility of the three item bundle (55) and the part-worth utility of the price DM 7.50 (20). Analogously, we derive the overall utilities for the other stimuli. From Table 4 we see that the consumer continues to prefer the three item bundle at a price of DM 9.00 over the other stimuli. We are now interested in the price which fulfills equation (2). At DM 10.50 the consumer is indifferent to the bundle of all the three items and the bundle of two items such as Big Mac and French Fries. Hence, as long as the price is less than DM 10.50, the consumer will prefer the bundle of three items. Therefore, the derived reservation price of the three item bundle for this consumer is DM 10.49. Assuming that the consumer will buy the bundle which maximizes his or her utility, the consumer in Table 4 would still buy the three item bundle at DM 10.49.

Table 4. Overall utility of the seven stimuli for one consumer[a]

Bundle/Item Type	Price (DM)	Overall Utility
French Fries, Big Mac and Drink	7.50	75
French Fries, Big Mac and Drink	9.00	70
French Fries, Big Mac and Drink	10.50	55
Big Mac and French Fries	7.50	55
Big Mac	5.00	53
French Fries	2.50	50
Drink	2.50	35

[a] Based on Table 3. For example, overall utility of 75 is the sum of the utility of 55 for the three item bundle and the utility of 20 for the price level of DM 7.50.

Hence, by assessing the individual part-worth utilities via a conjoint analysis-based procedure, we can now derive the individual reservation prices for the three item bundle and the demand for different items. Based on this information, we can maximize the firm's profit. The profit function is given by following equation:

$$\Pi = \sum_{n=1}^{5} (p_n - c_n)D_n \ , \qquad (3)$$

where $\Pi =$ overall profit,

$p_n =$ price of item n,

$c_n =$ cost of item n,

$D_n =$ demand of item n.

The term $(p_n - c_n)$ denotes the contribution of item n. As we assume that each consumer buys only one unit of an item, the term $(p_n - c_n)$ can be interpreted as the contribution of each consumer to the firm's profit.

4 Application of the Proposed Approach in the Price Bundling Context

As an application of the proposed model, we consider the problem of finding the optimal price for the three item bundle (Big Mac and French Fries and Drink) reported in Table 2. We assume that every respondent makes a purchase and the option of not buying any of the bundle/item types is not available. This is a reasonable assumption, for we asked the customers in the restaurant itself. However, we note that the model described in Section 3 can easily be used to examine cases in which consumers have the option not to purchase.

We assume that only the price of the three item bundle is varied, i.e., the other prices are kept constant at their market prices. The stimuli described in Table 2 represent actual products offered by a major fast food chain in Germany. We conducted the data collection in a fast food restaurant in a major city in Germany. We interviewed 79 customers in the restaurant itself. Stimuli were presented with a laptop computer. After finishing the conjoint task, the customers filled out a short questionnaire that asked them to

1.) directly indicate their reservation prices for the five bundle/item types (listed in Table 1)
2.) rank the seven stimuli (listed in Table 2) and
3.) provide some demographic data such as age and occupation.

The interviews took approximately 10 to 15 minutes. In the first stage of the ACA program, we asked the customers to rank the levels of each attribute in terms of preference. The customer chose his or her most preferred level, next most preferred level, and so on. Since rank ordering for the price attribute is obvious, we did not include the price variables in this stage. In the second stage, we asked the consumers the importance of each attribute (Johnson, 1987). In the third stage, we asked subjects to choose between different pairs of multi-attribute stimuli. The number of these graded paired comparisons (PQ) is determined by the following built-in formula:

$$PQ = 3\,(N - n - 1) - N, \tag{4}$$

where N denotes the total number of levels and n denotes the number of attributes. Based on this information, the ACA procedure estimated the individual part-worth utilities using an OLS regression updating procedure to adjust the tentative part-worths in the first two stages to the customer's judgments in the third stage.

Based on the individual part-worth utilities, we can derive the individual reservation prices for the three item bundle as described earlier. Next, we can estimate the demand for the n items that can be used to maximize the firm's profit. Table 5 reports the number of buyers of different items and bundle types at various prices of the three-item bundle and the corresponding profits.

Table 5. Number of buyers for different bundles/items at various prices of the three item bundle and corresponding total profits

Price of the Three Item Bundle (in DM)	Number of Buyers of Different Bundles/Items					Total Profits (in DM)
	3 Item Bundle	Big Mac and Fries	Big Mac	Fries	Drink	
7.50	73	1	5	0	0	351.25
8.10	65	6	7	0	1	389.25
8.70	52	15	9	1	2	411.90
9.30[a]	42	23	11	1	2	429.10[a]
9.50	29	34	12	1	3	416.00
9.90	17	43	13	1	5	399.05
10.30	14	44	13	3	5	392.70
10.70	8	50	13	3	5	383.60

[a] Optimal price and maximized profit.

From Table 5, we see that the optimal price for the three-item bundle is DM 9.30. At this price, 42 customers buy the three-item bundle, 23 customers buy the two-item bundle (Big Mac and French Fries), 11 customers buy a Big Mac, two customers buy Fries and one customer buys a drink. This bundle pricing leads to a total profit of DM 429.10. Figure 1 shows the share of preferences for different bundle types at the various prices of the three-item bundle. We see in Figure 1 that many customers switch from the three-item bundle to the two-item bundle at a three-item bundle price above DM 9.30. Therefore, the company should be aware that setting a price higher than DM 9.30 will considerably reduce its profits. Because we also asked customers directly for their reservation prices for the three item bundle, we can compare these prices with the reservation prices that we derived via conjoint analysis. We expect that many subjects will underestimate the reservation price with the direct method since they may perceive their role as conscientious buyers to be to help keep prices down.

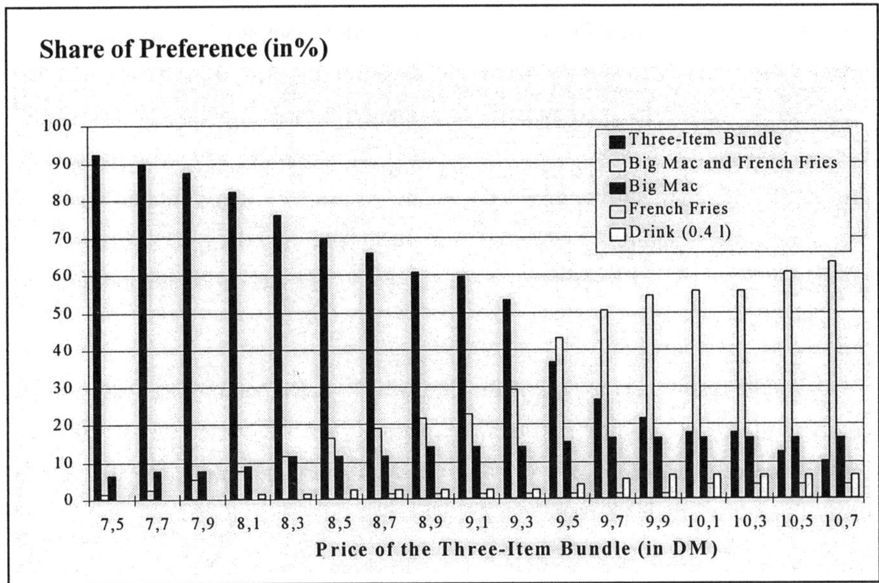

Fig. 1. Share of preferences for different bundles of items at various prices of the three-item bundle

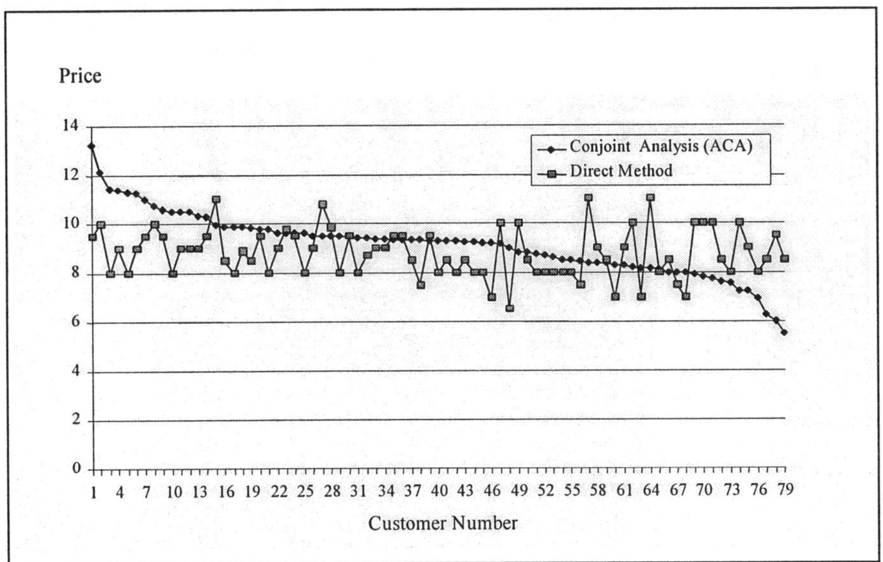

Fig. 2. Comparison between the individual reservation prices for the three-item bundle obtained via ACA and direct method

We see in Figure 2 that the majority of the respondents (65 percent) expressed a lower reservation price for the three-item bundle, compared to the reservation price determined via conjoint analysis. Interestingly, over 30 percent stated a higher reservation price. One possible explanation for this may be that the price is viewed in isolation by using the direct method. This implies, for example, that the prices of the competitive products (such as the two-item bundle) are not considered when stating the reservation price for the three-item bundle. Figure 2 presents the individual reservation prices obtained via the direct method and via conjoint analysis.

Using the reservation prices via the direct method and conjoint analysis we can compare the corresponding profits. We assume a specific variable cost structure (e.g. 30 percent of the market price). Based on this assumption, we derive the profit functions which are presented in Figure 3.

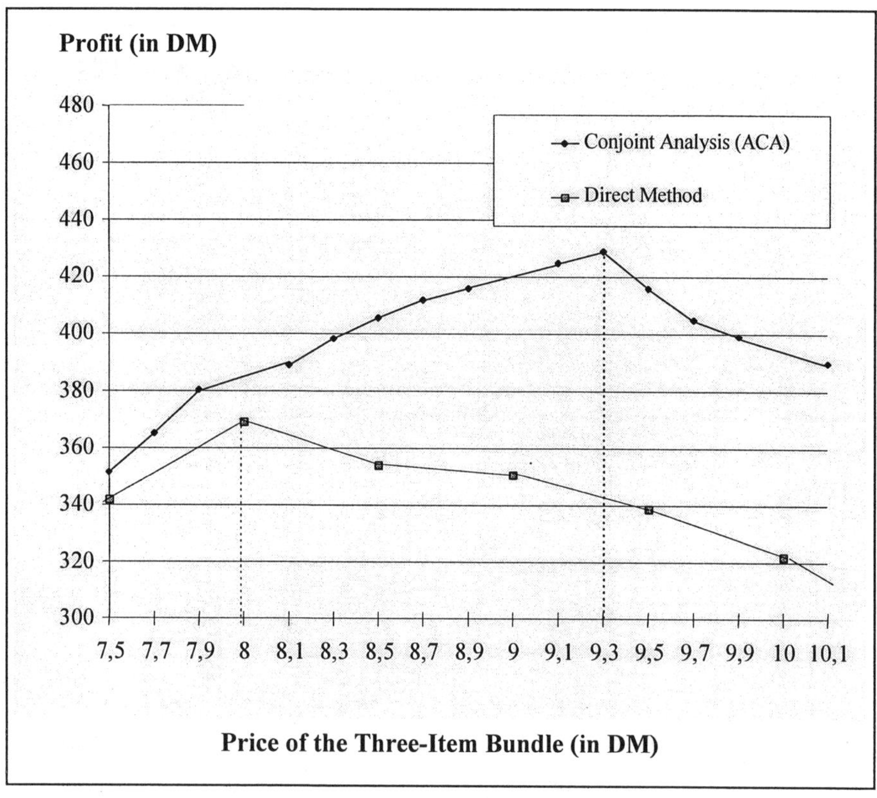

Fig. 3. Profit comparison (ACA vs. direct method)

This Figure shows that the optimal bundle price using the direct method is DM 8.00 and the corresponding profit is DM 362.25. By using conjoint analysis, we derive the optimal bundle price at DM 9.30 which leads to a profit of DM 429.10 - a nearly 20 percent increase. The empirical results are summarized in Table 6.

Table 6. Comparison of empirical results based on conjoint analysis and direct method

	Procedure		
Criteria	**Conjoint Analysis**	**Direct Method**	**Difference**
Mean Reservation Price[a]	9.15	8.78	+ 0.37[b]
Optimal Bundle Price[a]	9.30	8.00	+ 1.30[b]
Corresponding Share of Preferences	53.2%	88.6%	- 35.4%[b]
Corresponding Profit (Variable Cost: 30%)[a]	429.10	369.00	+ 60.10[b]

[a] All Prices in DM

[b] Significant at 5% level

In the questionnaire, we also asked the customers to rank the seven stimuli presented in Table 2. These ranking data were used to check how well the estimated individual part-worth utilities from the conjoint analysis-based procedure predict these ranks. Three measures were used to check these relationships. We calculated the Spearman correlation coefficient, and the percentage of correct first-choice hits. A third validity criterion that we used was Kendall's Tau, which is defined as follows (Siegel, 1956):

$$\text{Kendall's Tau} = \frac{N_c - N_{nc}}{N}, \tag{5}$$

where N = Total number of paired comparisons,

N_c = Number of correctly predicted paired comparisons,

N_{nc} = Number of incorrectly predicted paired comparisons.

Table 7 provides a summary of the three validity criteria. We see that these indicators suggest a good fit. The mean Spearman correlation coefficient is 0.974. This coefficient is used to assess how well the rank order data (derived from the part-worth utilities of the ACA procedure) fit the rank order data of the direct method. Kendall's Tau is a measure for the internal validity of the estimated part-worth utilities. It is bounded by 0 and 1. In our study the mean Kendall's Tau is 0.899. This indicates a good fit compared to other studies (Srinivasan, Jain and Malhotra, 1981; Mason, 1990). The first-choice hit is another measure for the internal validity of the estimated part-worth. In our study, conjoint analysis correctly predicted 71 out of 79 consumers' first choices. This also indicates a good fit.

Table 7. Tests of validity for estimated part-worth utilities

Validity Criteria	Results
Spearman Correlation Coefficient	0.974 [a]
Kendall's Tau	0.899[a]
First-Choice Hit	0.898[b]

[a] Indicates the mean of the 79 customers

[b] 71 out of 79 gave the same first choice via conjoint analysis predicted and self-explicated ranking.

5 Conclusions

The objective of our paper was to propose a conjoint analysis-based procedure to measure reservation price and to optimally price product bundles. In contrast to previous studies, we characterized the bundle type and the price as two relevant attributes in a conjoint model. The conjoint data are used to infer individual reservation prices. We also asked consumers to indicate their reservation prices directly for seven stimuli so that we could compare the results of the conjoint analysis-based procedure with the results of the direct method. An application in the fast food industry provides several interesting findings. First, we find that the majority of the respondents underestimates the bundle reservation price when asked directly. This underestimation yields an incorrectly estimated bundle demand function. Prices derived from this demand function lead to sub-optimal choices. Therefore, the firm earns a sub-optimal profit (in our application nearly 20 percent). Second, assessing reservation prices via a conjoint based procedure leads to more realistic results, because the the price is not evaluated in isolation. In other words the conjoint analysis addresses the issue of trade-offs in a systematic way. In summary, estimating reservation prices via the conjoint analysis-based procedure leads to superior results in predicting respondent preferences for various bundle types, measuring reservation prices; and achieving higher profits.

There are also some limitations in our study that may affect the degree to which our results can be generalized. We assumed that the single item prices and the two-item bundle price are kept constant, i.e., only the three-item bundle price varies (over three price levels). An extension of this study would be to vary the prices of all items simultaneously to assess the reservation prices for each item. Another limitation in our application is our holding the number of stimuli to seven. Therefore, we do not have a strictly orthogonal design. Further studies should focus on designs which address these problems. Furthermore, our model does not consider competition. This provides an opportunity for interesting extensions in future research on bundling.

References

Adams, W.J. and J. L. Yellen (1976). “Commodity Bundling and the Burden of Monopoly.” Quarterly Journal of Economics, Vol. 90 (August), 475-498.

Agarwal, M.K. and P. E. Green (1991). “Adaptive Conjoint Analysis versus self-explicated models: Some empirical results.” International Journal of Research in Marketing, Vol. 8 (June), 141-146.

Anderson, J.C. and J. A. Narus (1995). “Capturing the Value of Supplementary Services.” Harvard Business Review, Vol. 73 (January/February), 107-117.

Burstein, M.L. (1960). “The Economics of Tie-In Sales.” Review of Economics and Statistics, Vol. 42 (February), 68-73.

Carbajo, J., D. de Meza and D. J. Seidmann (1990). “A Strategic Motivation for Commodity Bundling.” The Journal of Industrial Economics, Vol. 38 (March), 283-298.

Drumwright, M. E. (1992). “A Demonstration of Anomalies in Evaluations of Bundling.” Marketing Letters, Vol. 3 (October), 311-321.

Eppen, G.D., W. A. Hanson and K. R. Martin (1991). “Bundling - New Products, New Markets, Low Risk.” Sloan Management Review, Vol. 32 (Summer), 7-14.

Fuerderer, R., A. Huchzermeier and L. Schrage (1994). “Stochastic Option Bundling and Bundle Pricing.” Working Paper, WHU Koblenz, Germany.

Gaeth, G. J., I. P. Levin, G. Chakraborty and A. M. Levin (1990). "Consumer Evaluation of Multi-Product Bundles: An Information Integration Analysis." Marketing Letters, Vol. 2 (January), 47-57.

Green, P. E. and V. Srinivasan (1990). "Conjoint Analysis in Marketing: New Developments With Implications for Research and Practice." Journal of Marketing, Vol. 54 (October), 3-19.

Green, P. E., A. M. Krieger and M. K. Agarwal (1991). "Adaptive Conjoint Analysis: Some Caveats and Suggestions." Journal of Marketing Research, Vol. 28 (May), 215-222.

Guiltinan, Joseph P. (1987). "The Price Bundling of Services: A Normative Framework." Journal of Marketing, Vol. 51 (April), 74-85.

Hanson, W. A. and K. R. Martin (1990). "Optimal bundle pricing." Management Science, Vol. 36 (February), 155-74.

Hayes, B. (1987). "Competition and Two-Part Tariffs." Journal of Business, Vol. 60 (January), 41-54.

Johnson, R. M. (1987). "Adaptive Conjoint Analysis." Sawtooth Software Conference on Perceptual Mapping, Conjoint Analysis, and Computer Interviewing. Ketchum, ID: Sawtooth Software, Inc., 253-265.

Johnson, R. M. (1991). "Comment on Adaptive Conjoint Analysis: Some Caveats and Suggestions." Journal of Marketing Research, Vol. 28 (May), 223-225.

Kaicker, A., W. O. Bearden and K.C. Manning (1995). "Component versus Bundle Pricing: The Role of Selling Price Deviations from Price Expectations." Journal of Business Research, Vol. 33 (June/July), 231-239.

Kenney, R. W. and B. Klein (1983). "The Economics of Block Booking." Journal of Law and Economics, Vol. 26 (October), 497-540.

Kohli, R. and V. Mahajan (1991). "A Reservation-Price Model for Optimal Pricing of Multiattribute Products in Conjoint Analysis." Journal of Marketing Research, Vol. 28 (August), 347-354.

Kohli, R. and H. Park (1994). “Coordinating Buyer-Seller Transactions Across Multiple Products.” Management Science, Vol. 40 (September), 1145-1150.

Lawless, M. W. (1991). “Commodity Bundling for Competitive Advantage: Strategic Implications.” Journal of Management Studies, Vol. 28 (May), 267-280.

Lewbel, A. (1985). “Bundling of Substitutes or Complements.” International Journal of Industrial Organization, Vol. 3 (September), 101-107.

Liebowitz, S. J. (1983). “Tie-in Sales and Price Discrimination.” Economic Inquiry, Vol. 21 (July), 387-399.

Mason, C. H. (1990). “New Product Entries and Product Class Demand.” Marketing Science, Vol. 9 (Winter), 58-73.

Monroe, K. B. (1990). Pricing: Making Profitable Decisions. New York: McGraw-Hill.

Morton, J. (1989). “The Economics of Price.” In: Daniel T. Seymour (ed.). The Pricing Decision. Illinois: Probus Publishing Company.

Palfrey, T. R. (1983). “Bundling Decisions by A Multiproduct Monopolist with Incomplete Information.” Econometrica, Vol. 51 (March), 463-483.

Paroush, J. and Y. C. Peles (1981). “A Combined Monopoly and Optimal Packaging.” European Economic Review, Vol. 15 (March), 373-383.

Porter, M. E. (1986). Wettbewerbsvorteile. Wiesbaden: Campus Verlag.

Schmalensee, R. (1984). “Gaussian Demand and Commodity Bundling.” Journal of Business, Vol. 57 (January), 211-30.

Siegel, S. (1956). Nonparametric Statistics for the Behavioral Sciences. New York: McGraw-Hill.

Simon, H. (1989). Price Management. Amsterdam: North-Holland.

Simon, H. (1992). “Preisbündelung.” Zeitschrift für Betriebswirtschaft, Vol. 62 (November), 1213-1235.

Srinivasan, V., Jain, A. K. and N. K. Malhotra (1981). "Improving Predictive Power of Conjoint Analysis by Using Constrained Parameter Estimation." Research Paper No. 621, Graduate School of Business, Stanford University.

Stigler, G. J. (1968). A Note on Block Booking. The Organization of Industry. Homewood, IL: Richard D. Irwin, Inc.

Venkatesh, R. and V. Mahajan (1993). "A Probabilisitic Approach to Pricing a Bundle of Products or Services." Journal of Marketing Research, Vol. 30 (November), 494-508.

Warhit, E. (1980). "The Economics of Tie-in Sales." Atlantic Economic Journal, Vol. 8 (December), 81-88.

Whinston, M. D. (1990). "Tying, Foreclosure, and Exclusion." American Economic Review, Vol. 80 (September), 837-859.

Wittink, D. R., M. Vriens and W. Burhenne (1994). "Commercial Use of Conjoint Analysis in Europe: Results and Critical Reflections." International Journal of Research in Marketing, Vol. 11 (March), 41-52.

Yadav, M. S. and K. B. Monroe (1993). "How Buyers Perceive Savings in a Bundle Price: An Examination of a Bundle's Transaction Value." Journal of Marketing Research, Vol. 30 (August), 350-358.

Yadav, M. S. (1994). "How Buyers Evaluate Product Bundles: A Model of Anchoring and Adjustment." Journal of Consumer Research, Vol. 21 (September), 342-353.

Part 3: Behavioral Aspects

Consumers Prior Purchase Intentions and their Evaluation of Savings on Product Bundles

Rajneesh Suri[1], and Kent B. Monroe[2]

[1] **Rajneesh Suri**, Drexel University, Philadelphia, Pennsylvania.
[2] **Kent B. Monroe**, University of Illinois, Urbana-Champaign, Illinois.

1 Introduction

Bundling of more than one product has been a strategy used by retailers and manufacturers for decades in both consumer and industrial markets (Nagle 1984). The hardware and software packages offered by computer manufacturers, vacation packages offered by the travel agencies, a shaving foam sold along with razors at some retailers are just a few commercial examples of bundling. While offering such product bundles, marketers usually use one of two forms of product bundling—either a pure bundle or a mixed bundle. When following a pure bundling strategy, the marketer offers a combination of certain products or services only in bundled form. However, when a marketer gives an option to consumers to buy these products or services separately or as a bundle, the bundling strategy is called mixed bundling. Given the popularity and the effectiveness of mixed bundling in marketing, this research will focus only on this type of bundling strategy.

There is extensive research on product bundling in the economics literature. However, given the macro orientation of this research and its limited applicability to marketing practice, marketing academics have recently started to aggressively pursue research in product bundling. The focus of some of the consumer behavior studies like those by Yadav (1990; 1994) and Yadav and Monroe (1993) have been to understand the process by which consumers evaluate product bundles. This current project extends this research stream in consumer behavior by understanding the effects of a contextual factor like consumers' prior purchase intentions on their evaluation of product bundles.

Specifically, Yadav and Monroe's (1993) study explored the effects of savings information provided by a bundle on consumers' evaluation of total savings on the bundle. In their study subjects evaluated a two-item luggage bundle consisting of a "Garment bag" and a "Pullman." They found that consumers' perceptions of total savings on a bundle is a sum of a) perceived saving on individual items if purchased separately, and b) the additional savings when the items in a bundle are bought together as a set. A closer look at the experimental procedures used in this study indicate that before evaluating the product bundle, subjects were instructed to (Yadav 1990):

> Imagine that you are interested in purchasing the following two luggage items 1) a Garment Bag and 2) a Pullman. **Also, assume that you wish to have matching luggage items and would therefore prefer to purchase these two luggage items together as a set** (p.398).

The above manipulation could be argued to provide subjects with only the option of having a prior intention to purchase both items in the bundle. One can argue however, that some consumers, while evaluating a similar two-item bundle, may not necessarily have intentions to buy both items in the bundle. Instead they may have prior intentions to:

> purchase only one of the items in a two-item bundle, or
> purchase neither item in the bundle.

We could think of these prior intentions to buy items in a bundle as consumers' purchase plans (Suri and Monroe 1995). These purchase plans could be either mental or physical notes kept by consumers and used by them during their shopping trips. Given the above possible additional scenarios it can be argued that the instructions given to subjects in Yadav's (1990; Yadav and Monroe 1993) study, merely accounted for one type of prior purchase intention. Hence, the *question remains as to how consumers will evaluate the total savings on a two-item bundle if they had prior intentions to buy only one or none of the items in a given bundle.*

2 Conceptual Development

Thaler (1985) proposed that the overall utility of single item transactions could be decomposed into acquisition utility and transaction utility. Even though Thaler developed the construct of transaction and acquisition utility for single item transactions, Yadav (1990) extended the same for transactions involving multiple items as in the case of product bundles. Briefly, Thaler (1985) defined acquisition utility (acquisition value in Monroe 1990; also Yadav 1990) as the "value of the good received compared to the outlay" (p.205). Extending this construct to bundle offers, Yadav (1990) suggested that acquisition value for a bundle will focus on two evaluations: 1) the processing of non-price information about bundle items to assess the worth of what is received, and 2) the processing of price information to assess the sacrifice it represents.

Similarly, Thaler (1985) argued that transaction utility (or transaction value in Monroe 1990: also Yadav 1990) depends "solely on the perceived merit of the deal" (p.205) or "the perception of savings in a transaction" (Yadav 1990, p.124). Though both transaction and acquisition utilities (values) affect the formation of overall utility of a bundle, Yadav and Monroe's (1993; also Yadav 1990) study limited its exploration to transaction utility only and so will this study. To understand the concept of transaction value (transaction utility) for bundles let us consider a simple example of a bundle consisting of a garment and a duffel bag (Table 1).

Table 1. Bundle price scenario

Item	Regular Price	Sale Price	Perceived Savings
Garment Bag	\$120	\$100	\$20
Duffel Bag	\$60	\$50	\$10
Total	\$180	\$150	\$30

OR buy both as a set for \$140

Following Yadav's (1990, p.126) model of transaction value for product bundles, consumers will determine three types of savings or transaction values for this bundle offer:

Total transaction value: The consumers' perception of savings associated with purchasing the bundle.

Bundle transaction value: The perception of savings associated with bundle items as a set versus purchasing the bundle items separately ($100 + $50 - $140 = $10).

Item transaction value: The perception of savings associated with the amount required to purchase all the bundle items separately ($20 + $10 = $30).

Using Thaler's (1985) argument that segregation of multiple gains instead of integration (i.e., perceiving multiple gains separately rather than jointly) leads to a higher perception of overall gains, Yadav and Monroe (1993) showed that consumers may prefer this method of framing savings in product bundles as well. The empirical results of their study showed that the total transaction value for a bundle was a sum of a) the item transaction value for the two items in the bundle, and b) the bundle transaction value. Similar support for the segregation of gains was also shown in the study by Chakravarti et al. (1994). This later study showed that a target bundle was evaluated more favorably and chosen more often when its components were presented in segregated (separate price tags) versus consolidated (single equivalent tag) fashion, a notion consistent with mental accounting predictions (Thaler 1985).

Given our argument that Yadav and Monroe's (1993) study had implicitly manipulated subjects' prior intentions to purchase both items in a bundle, the question remains whether the results from Yadav and Monroe's study will hold for situations where consumers did not have similar prior purchase intentions. In other words, will consumers add bundle transaction value and item transaction value to arrive at the total transaction value for a bundle even when they had prior intentions to purchase only one or none of the items in a bundle? In the following section we develop a conceptualization to support the predictions for the latter two situations.

3 Prior Intentions to Purchase only One or None of the Items in a Bundle

Yadav (1990, 1994) argues that although a bundle may have only a few items, the amount of information available for processing can be substantial and it is likely that buyers will look for ways to simplify the evaluation task. The results from the study by Yadav (1990, 1994) showed that one of the ways consumers simplified their bundle evaluation was by following a sequential "step by step" structure consistent with the anchoring and adjustment heuristic proposed by Tversky and Kahneman (1974). This heuristic suggests that an arbitrarily chosen reference point will significantly affect the total savings estimate for a bundle.

Lopes (1982) proposed a similar model of anchoring and adjustment process but identified three stages during the process, 1) scanning, 2) anchor selection, and 3) anchoring and adjustment. Lopes suggests that it is during the anchor selection stage that buyers initiate the evaluation process and select one piece of information perceived to be most important for the evaluation task. The evaluation of the selected piece of information serves as a starting point for the evaluation of the bundle. The remaining pieces of information are then evaluated in decreasing order of their importance. If a consumer is given the bundling scenario in Table 1, the question is what information will the consumer use as an anchor to evaluate the total transaction value of a bundle? To answer this question the results from Yadav and Monroe's (1993) study provide some assistance. This study found that while evaluating a bundle consumers are more likely to attend to savings on the bundle first and then the savings on individual items. Given this empirical result it can be argued that consumers will use bundle savings (bundle transaction value) to initiate the evaluation of total savings on the bundle. This initial evaluation will then be adjusted depending upon the evaluation of savings information on individual items that will be evaluated in a decreasing order of their importance.

Given that a consumer has plans for buying only one or none of the items in a bundle it can be argued that the information about savings on individual items could be perceived as irrelevant or less important piece(s) of information. Past research suggests that while evaluating different pieces of information consumers ignore redundant or less important pieces of information in their attempts to minimize their cognitive effort (Beach and Mitchell 1978; Payne, Bettman and Johnson 1988). Similarly, Lopes (1982) suggests that for simple tasks, decision makers may anchor on the most important piece of evidence and then make only minor (small) adjustments on the basis of an aggregate evaluation of the remaining pieces of evidence.

Essentially these conceptualizations suggest that the adjustment of the total savings on the bundle due to the evaluation of the item savings will be insufficient and the final evaluation of the total savings will be primarily a function of the savings on the bundle. This argument is consistent with Tversky and Kahneman's (1974) suggestion that the evaluation of an anchor under the anchoring and adjustment heuristic tends to be insufficient, and the final evaluation is biased in the direction of the initial anchor evaluation. Empirical support for Kahneman and Tversky's argument could be observed in spouse's evaluation of new products (Davis et al. 1986), realtors' determination of fair market value of residential properties (Northcraft and Neale 1987), and in auditors' evaluation of fraud estimates (Joyce and Biddle 1981).

Given the above evidence and our conceptualization, it can argued that for a consumer with prior intentions to buy only one or none of the items in a bundle, the total transaction value for the bundle will be primarily influenced by the bundle transaction value. The less important or redundant information about the item transaction value will have little or no impact on the total transaction value. Hence it is hypothesized that:

> ***Hypothesis:*** *Given that a consumer has prior intentions to buy one or none of the items in a two item bundle, the total transaction value for the bundle will be influenced by the bundle transaction value with little or no impact of the item's transaction value.*

An experiment was conducted to test the above hypothesis. In addition to testing this hypothesis, the experiment also evaluated different processes by which consumers may use the given price information to calculate the total savings (total transaction value) for the bundle.

4 Research Method

To examine the effects of prior intentions to purchase one or none of the items in a bundle, 205 undergraduate students (41% male and 59% female) at a large U.S. state university were recruited for the study and given extra credit for their participation. Similar to product stimuli used in studies by Yadav (1990) and Yadav and Monroe (1993), this study also used two luggage items as a product bundle. The luggage items were a garment bag and a duffel bag. These items were selected because pretests showed subjects to be reasonably knowledgeable about these products as well as their prices.

Design

To understand the effects of prior purchase intention on the total transaction value, three bundle prices along with two savings on the unplanned purchase were used in a 3 (prior intentions to purchase) x 3 (bundle price) x 2 (savings on unplanned purchase) between subjects design. Three vignettes were used to manipulate the prior intentions to purchase either one or none of the items in a two-item bundle (see Appendix 1). A pretest used to determine regular prices for the two luggage items resulted in our decision to use $120 as the regular price for the garment bag and $60 for the duffel bag. The three bundle prices used in this study were $120, $130, and $140 and the saving on the unplanned purchase was kept at either $0 or $10. The savings on the planned purchase was $20 for the garment bag (regular price $120) and $10 for the duffel bag (regular price $60). The six different combinations of regular and sale price at a bundle price of $140 are shown in Table 2.

Table 2. Different bundle price scenarios at a bundle price

Prior Intentions	**Item**	**Regular Price**	**Sale Price**
Buy Garment Bag only			
Unplanned Savings: $ 10	Garment Bag	$ 120	$ 100
	Duffel Bag	$ 60	$ 50
Unplanned Savings: $ 0	Garment Bag	$ 120	$ 100
	Duffel bag	$ 60	$ 60
Buy Duffel bag Only			
Unplanned Savings: $ 10	Garment Bag	$ 120	$ 110
	Duffel Bag	$ 60	$ 50
Unplanned Savings: $ 0	Garment Bag	$ 120	$ 120
	Duffel Bag	$ 60	$ 60
Buy Neither Item			
Unplanned Savings: $ 10	Garment Bag	$ 120	$ 110
	Duffel Bag	$ 60	$ 50
Unplanned Savings: $ 0	Garment Bag	$ 120	$ 120
	Duffel Bag	$ 60	$ 60

OR buy both as a set for $140

Dependent Measures

Following Yadav and Monroe's (1993) study eleven items were used to operationalize the three dependent measures; total transaction value (3 items), bundle transaction value (4 items) and item transaction value (4 items). In addition to these measures, subjects were also asked an open-ended question. This question asked subjects to write down the steps they had followed to evaluate the overall savings on the bundle offer. The purpose of this open-ended question was to determine processes by which consumers evaluated savings on a two-item bundle offer.

Procedure

The response booklet first introduced subjects to one of three purchase vignettes. The following section in the response booklet introduced subjects to the description of the product bundle. This description included a list of features about the two products, their regular and sale prices and the bundle price for buying the two products as a set. After examining the product description subjects responded to the three dependent measures and the open ended question. Finally, in the last section subjects indicated their estimate of monetary saving on buying the product bundle. This section also collected classification information like age and gender. Two hundred and five participants in this study resulted in final cell sizes ranging from 11 to 12 for all cells. The test for homogeneity of variance showed no significant effect of imbalanced cell design on variance on dependent measures in the various cells.

5 Results

Reliability and Manipulation Checks

Reliability of the three dependent measures was satisfactory, with the Cronbach's alpha greater than 0.80 for all three constructs (total transaction value $\alpha=0.82$; bundle transaction value $\alpha=0.86$; item transaction value $\alpha=0.95$). Using monetary savings on the bundle as a measure to test the manipulation of the bundle price showed that the mean value on this measure was significantly different for the three bundle prices ($F(2, 202) = 38.8$, $p<0.00$). The mean value of monetary saving for the bundle was \$45.50 at a bundle price of \$120, \$37 at a bundle price of \$130, and \$27.70 at a bundle price of \$140.

The Effect of Prior Purchase Intentions

The MANOVA using three dependent measures and the two factors—prior purchase intentions and bundle price—showed significant effect when the savings on unplanned purchase was \$0 (prior purchase intention: Pillais $F(6, 182) = 16.38$, $p<0.00$.eta2 =0.35; bundle price Pillais $F(6, 182) = 3.80$, $p<0.00$, eta2 =0.12) as well as when it was \$10 (prior purchase intention: Pillais $F(6, 188) = 3.97$, $p<0.00$.eta2 =0.19; bundle price Pillais $F(6, 188) = 5.83$, $p<0.00$, eta2 =0.16). To understand these results further and to determine support for the hypotheses we evaluated the effects of prior purchase intentions on the three dependent measures at each level of savings on the unplanned purchase item. Two approaches were used to perform this analysis. The first approach involved a comparison of cell means for the three bundle prices for each prior purchase intention and the second approach used regression analyses.

Comparison of cell means for the three purchase plans showed support for the notion that a reduction in bundle price resulted in an increase in the total transaction value as well as the bundle transaction value (see Table 3). The regression analysis using total transaction value as the dependent measure and bundle transaction value and item transaction value as the two independent measures are shown in Table 4. The results from the regression analysis showed that when subjects had prior intentions to buy one or none of the items in a bundle, the bundle transaction value was a significant predictor for the total transaction value on the bundle. On the other hand the item transaction value had no significant effect on the total transaction value.

Table 3. Impact of prior purchase intentions on the dependent measures

Prior Intentions to Buy	Unplanned Item Savings	Dependent Measure	Bundle Price $120	Bundle Price $130	Bundle Price $140	F-value
Garment Bag	$0	BTV	5.4	5.2	4.3	3.4a
		TTV	5.7	5.1	4.6	2.8b
		ITV	5.1	5.3	5.1	1.0
	$10	BTV	5.2	4.8	4.1	4.1a
		TTV	5.5	5.3	4.8	2.4
		ITV	3.5	4.4	4.2	2.2
Duffel Bag	$0	BTV	5.5	5.2	4.6	2.3
		TTV	5.7	5.1	4.4	4.2a
		ITV	3.7	4.5	4.5	2.2
	$10	BTV	5.2	4.8	4.0	5.5a
		TTV	5.6	5.2	4.2	6.2a
		ITV	3.0	3.1	3.5	0.6
Neither Item	$0	BTV	5.6	5.7	4.3	6.3a
		TTV	5.2	5.7	4.5	3.2a
		ITV	1.7	1.7	1.7	0.0
	$10	BTV	5.4	5.2	4.3	2.6b
		TTV	5.3	5.2	4.6	0.7
		ITV	2.5	3.0	3.5	2.5b

BTV = Bundle transaction value; TTV = Total transaction value; ITV = Item transaction value. The values for the dependent measures represents cell means on a 7-point scale where 1 is a low and 7 is a high. Significance levels for the F-value are indicated as a = $p < 0.05$; b = $p < 0.10$; all other values were not significant.

Table 4. Regression coefficients indicating the impact of bundle transaction value and item transaction value on the total transaction value

Independent Variable		Garment Bag Only			Prior Intentions to Buy Duffel Bag Only			Neither		
	Unplanned Savings:	$0	$10	Total	$0	$10	Total	$0	$10	Total
BTV		0.78 (6.6) a	0.69 (6.1) a	0.74 (8.5) a	0.80 (7.2) a	0.70 (4.9) a	0.74 (8.8) a	0.90 (8.7) a	0.92 (13.7)a	0.90 (16.1)a
ITV		-0.02 (-.15)	0.21 (2.1)	-0.1 (-.85)	-0.06 (-.56)	0.03 (.20)	0.03 (.40)	0.13 (1.3)	0.08 (1.16)	0.11 (2.0)

The values in parenthesis are t-statistics. Significant level for the t-statistics is indicated by a=p<0.05. BTV = Bundle transaction value; TTV = Total transaction value. "Total" represents results of the regression analysis after combining responses for conditions when unplanned savings were $0 and $10.

Process Used to Evaluate Total Savings on the Bundle

An open-ended question in this study was used to understand the process used by respondents to evaluate the total savings on a bundle offer. Since our objective was to understand the effect of prior intentions to purchase items in a bundle on the processing of price information in the bundle, the data were combined across the other two factors and analyzed for the three prior purchase intentions only.

A review of the responses indicated that one of fourteen different processes was used to evaluate the total savings on the bundle. These fourteen processes are listed in Appendix 2. Two independent judges coded each response as one of these fourteen processes. There was agreement between the judges on 90% of the coding of the responses and disagreements were resolved on the basis of a discussion between the judges.

Table 5. Three process categories used to evaluate the total transaction value

Category 1 (S.PA + S.PB) - B.P	Category 2 S.PB - (B.P - S.PA) or S.PA - (B.P - S.PB	Category 3 Others
Process # 1	Processes # 2 & 3	Processes # 4 through 14

These fourteen processes were then divided into a simpler classification consisting of three process categories (Table 5). The first category included the process that used bundle price and the sum of sale prices for the two bundle items to determine the total transaction value (process #1 in Appendix 2). Process category #2 included processes that evaluated the total transaction value by comparing the difference between the bundle price and the sale price of the planned purchase item with the sale price of the unplanned purchase (process #2 and #3 in Appendix 2). The third process category included all the remaining processes. A close look at the two process categories #1 and #2 indicates that both essentially result in same total monetary savings on the bundle. For instance, using the bundle scenario from Table 1 the total savings using process category #1 will result in a monetary savings of $10 [($100+$50)-$140] which is the same had the subjects used process category #2 [$50-($140-$100), or $100-($140-$50)]. A key difference between these two categories, however, is the anchor used to evaluate the sale price of the unplanned item. In process category#1 bundle price acts as the anchor while in process category #2, the difference between bundle price and the sale price of planned purchase item ($140-$50 or $140-$100) assumes the role of the anchor.

Table 6. Classification of responses into the three process categories

Prior Intentions to Buy	Category 1 (S.PA + S.PB) - B.P	Category 2 S.PB - (B.P - S.PA) or S.PA - (B.P - S.PB)	Category 3	Uncodable
Garment Bag	6	16	20	27
Duffel Bag	14	19	13	21
Neither	28	13	2	26

The summary of responses shown in Table 6, classified the responses into these three processes categories. Overall, the results in this table show that subjects' prior intentions to purchase significantly influenced the process used by them to evaluate the total savings on the bundle (overall Chi-Square (6) = 31.28, p<0.00; prior intention to buy neither item vs. prior intention to buy a garment bag, Chi-square (3) = 29.30, p<0.00; prior intention to buy neither item vs. prior intention to buy a duffel bag only, Chi-Square (3) =14.40, p<0.00). For subjects with prior intentions to purchase only one item in a bundle, process category #2 dominated the evaluation of the total savings on the bundle.

On the other hand, for subjects with prior intentions to purchase neither item, process category #1 dominated the process used to evaluate the total savings on the bundle. There was no significant difference in the process used to evaluate the savings when the prior intentions were to purchase either the garment bag or the duffel bag (Chi-Square (3) = 5.66, p>0.10). These results show that respondents with prior intentions to purchase only one item in a bundle preferred to evaluate the total savings on a bundle by comparing the difference between bundle price and the sale price of planned purchase item with the sale price of unplanned purchase item (i.e. process category 2). However, subjects with prior intentions to purchase neither item in a bundle evaluated the total savings on the bundle by comparing the bundle price with the sum of sale prices of the two bundle items, i.e., process category #1.

6 Discussion

The results of the regression analysis showed that when subjects had prior intentions to purchase either one or none of the items in a two-item bundle, the total transaction value of the bundle was significantly influenced by the bundle transaction value only. This result is consistent with that from Yadav and Monroe's (1993) study, which showed that while evaluating the overall savings on a bundle the savings on the bundle itself had a greater impact than the savings on individual items. Our current research extended the past research by showing that for consumers with prior intentions to purchase one or none of the items in a bundle the impact of savings on individual items on the total savings was even further reduced. Hence, given that a consumer has no prior intentions to buy all the items in a bundle, the bundle transaction value is a key predictor of the overall or total savings on the bundle.

Though we did not present any hypothesis about the process that could be used by consumers to determine the overall savings in a bundle, the open-ended question used in this study provided interesting insights into this process. The results showed that subjects with prior intentions to purchase none of the items in a bundle evaluated total savings on the bundle by anchoring on the bundle price and then comparing it with the sum of the sale prices of the two unplanned purchase items. On the other hand, subjects with prior intentions to purchase only one item in a bundle anchored on the difference between the bundle price and the sale price of the planned purchase item. These subjects then compared this anchor with the sale price of the unplanned item to arrive at their evaluation of the total savings on the bundle.

In essence the availability of a product bundle made subjects anchor on the bundle price which was then evaluated by comparing it with the monetary outlay required to buy the unplanned (unwanted) items in the bundle. This process was followed in the situation where subjects had no prior intention to buy any of the items in a bundle. In situations where subjects had prior intentions to buy one of the items in a bundle, the anchor was not the bundle price but rather the difference between the bundle price and the sale price of the planned purchase item. This anchor was then used to compare the monetary outlay required to buy the unplanned item in the bundle. However, in neither of these situations was the saving on individual items used to evaluate the total savings on the bundle. This conclusion is consistent with our conceptualization which suggests that for consumers with prior intentions to buy only one or none of the items in a bundle, the savings on individual items could be considered as redundant information which will not influence the evaluation of total savings on a bundle. Overall, the results from this study extend the research by Yadav and Monroe (1993) by showing that consumers' prior intentions to purchase or not to purchase certain items in a bundle could influence the evaluation of total savings on the bundle.

Appendix 1

Three Vignettes Used to Manipulate Prior Purchase Intentions

Prior intentions to buy a garment bag only

Vignette 1: Assume that you are interested in buying a garment bag and have been consulting different catalogs and visiting different stores. At your favorite store you notice that the garment bag that you are interested in is available at a sale price. The same store is also offering this garment bag as a bundle (a set)-packaged along with a duffel bag. However, please keep in mind that you have another duffel bag at home.

Prior intentions to buy duffel bag only

Vignette 2:Assume that you are interested in buying a duffel bag and have been consulting different catalogs and visiting different stores. At your favorite store you notice that the duffel bag that you are interested in is available at a sale price. The same store is also offering this duffel bag as a bundle (a set)- packaged along with a garment bag. However, please keep in mind that you have another garment bag at home.

Prior intentions to buy neither garment bag nor duffel bag

Vignette 3:Assume that you have been out shopping in the mall for a pair of shoes, when you come across a bundle offer for the purchase of a garment bag and a duffel bag as a bundle (set) at your favorite store. However, please keep in mind that you already have a duffel bag as well as a garment bag at home.

Appendix 2

Processes Used to Evaluate the Total Savings on a Bundle

Process 1: Total transaction value is perceived as a comparison of the individual items' sale prices with the bundle price.

Process 2: Total transaction value is perceived as a comparison of B's sale price with the additional amount one pays to buy the bundle over and above the sale price of A.

Process 3: Total transaction value is perceived as a comparison of A's sale price with the additional amount one pays to buy the bundle over and above the sale price of B.

<u>Process 4:</u> Total transaction value is perceived as a comparison of the individual items' regular prices to the bundle price.

<u>Process 5:</u> Total transaction value is a combination of the perceived additional savings on the bundle and the perceived savings on both items if bought separately.

<u>Process 6:</u> Total transaction value is perceived as a comparison of B's sale price with the additional amount one pays to buy the bundle over and above the regular price of A.

<u>Process 7:</u> Total transaction value is perceived as a comparison of A's sale price with the additional amount one pays to buy the bundle over and above the regular price of B.

<u>Process 8:</u> Total transaction value is perceived as a comparison of B's regular price with the additional amount one pays to buy the bundle over and above the regular price of A.

<u>Process 9:</u> Total transaction value is perceived as a comparison of B's regular price with the additional amount one pays to buy the bundle over and above the sale price of A.

<u>Process 10:</u> Total transaction value is perceived as a comparison of A's regular price with the additional amount one pays to buy the bundle over and above the sale price of B.

<u>Process 11:</u> Total transaction value is perceived as a comparison of A's regular price with the additional amount one pays to buy the bundle over and above the regular price of B.

<u>Process 12:</u> Total transaction value is perceived as a comparison of the individual items' regular prices to their sale prices.

<u>Process 13:</u> Total transaction value is a combination of the perceived additional savings on the bundle and the perceived item savings on A only.

<u>Process 14:</u> Total transaction value is the combination of the perceived additional savings on the bundle and the perceived item savings on B only.

References

Beach, L. R. and T. R. Mitchell (1978). "A Contingency Model for the Selection of Decision Strategies." Academy of Management Review (3), 439-449.

Chakravarti, Dipankar, Rajan Krish, Pallab Paul and Joydeep Srivastava (1994). "Segregated and Consolidated Presentation and Pricing of Product Bundles: Influences on Evaluation and Choice." Working Paper, University of Arizona, Tucson, Arizona 85721.

Davis, H. L., S. J. Hoch, and E. K. Easton Ragsdale (1986). "An Anchoring and Adjustment Model of Spousal Predictions." Journal of Consumer Research, 13 (June), 25-37.

Joyce, E. and G. C. Biddle (1981). "Anchoring and Adjustment in Probabilistic Inference in Auditing." Journal of Accounting Research, 19 (Spring), 120-145.

Lopes, L. L. (1982). "Toward a Procedural Theory of Judgement." Working Paper 17, Wisconsin Human Processing Program, Department of Psychology, University of Wisconsin, Madison.

Monroe, K. B. (1990). Pricing Making Profitable Decisions. New York, NY: McGraw-Hill Book Company.

Nagle, T. (1984). "Economic Foundations for Pricing." Journal of Business, 57 (No.1, Part 2), S3-26.

Northcraft, G. B. and M. A. Neale (1987). "Experts, Amateurs, and Real Estate: An Anchoring and Adjustment Perspective on Property Pricing Decisions." Organizational Behavior and Human Decision Processes, 39 (February), 84-97.

Payne, J. W., J. R. Bettman and E. J. Johnson (1988). "Adaptive Strategy Selection in Decision Making." Journal of Experimental Psychology: Learning, Memory and Cognition, (14), 534-552.

Suri, R. and K. B. Monroe (1995). "Effect of Consumers' Purchase Plans on the Evaluation of Bundle Offers." Advances in Consumer Research, Vol. 22, eds. Mita Sujan and Frank Kardes, Provo, UT: Association for Consumer Research, 588-593.

Thaler, R. (1985). "Mental Accounting and Consumer Choice." Marketing Science, 4 (Summer). 199-214.

Tversky, A. and D. Kahneman (1974). "Judgement under Uncertainty: Heuristics and Biases." Science, 185 (September), 1124-1131.

Yadav, M. S. (1990). "An Examination of How Buyers Subjectively Perceive and Evaluate Product Bundles." Unpublished doctoral dissertation, Department of Marketing, Virginia Polytechnic Institute and State University, Blacksburg, VA 24061.

Yadav, M. S. (1994). "How Buyers Evaluate Product Bundles: A Model of Anchoring and Adjustment." Journal of Consumer Research, 21 (September), 342-353.

Yadav, M. S. and K. B. Monroe (1993). "How Buyers Perceive Savings in a Bundle Price: An Examination of Bundle's Transaction Value." Journal of Marketing Research, 30 (August 1993), 350-358.

Buyers' Evaluations of Mixed Bundling Strategies in Price-Promoted Markets

Georg Wuebker[1], Vijay Mahajan[2], and Manjit S. Yadav[3]

[1] **Georg Wuebker**, SIMON, KUCHER & PARTNERS, Bonn, Germany.

[2] **Vijay Mahajan**, Department of Marketing, The University of Texas, Austin, Texas.

[3] **Manjit S. Yadav**, Department of Marketing, Texas A&M University, College Station, Texas.

1 Introduction

Price bundling, the joint offering of two or more items at a single price, is increasingly being viewed as strategically important for stimulating demand and increasing profit (Guiltinan, 1987; Simon, 1992; Dolan and Simon, 1996; Wuebker, 1998). The literature in this area has traditionally focused on sellers' motivations for implementing bundling strategies, such as price discrimination (Stigler, 1968; Adams and Yellen, 1976; Schmalensee, 1984), extension of monopoly power (Burstein, 1960; Whinston, 1990) or cost advantages (Kenney and Klein, 1983). In marketing, two main streams of research on bundling can be observed. One stream has focused on extending previous economic analyses, developing sophisticated models for optimizing the design and pricing of bundles (Hanson and Martin, 1990; Venkatesh and Mahajan, 1993, Fuerderer, 1996). A second research stream has focused on the behavioral aspects of bundling to understand consumers' evaluations of bundles (Drumwright, 1992; Gaeth et al., 1990; Yadav and Monroe, 1993; Yadav, 1994, 1995).

Bundling strategies are often accompanied by promotions on individual items. That is, firms promote their products individually as well as offer them as part of a package at a special price. Many examples can be given:

- In the computer retail industry, computers and printers are frequently price-promoted individually and are also available in a bundle or package at a special price.
- In the fast food industry, companies frequently price-promote value meals as well as individual items.
- Precision' Tune, a regional car workshop chain which frequently price-promotes its different tune-ups like maintenance or fuel injection tune-ups, also offers a set of tune-ups at a special price.
- American Airlines (AA) has its own vacation division, called AA Fly Away Vacation, which offers vacation packages like hotel accommodations plus airfare at a single price. Airline tickets are also frequently price promoted at deep discounts.

The firms' motivation for price-promoting the individual items is to enhance the sale of these individual items. However, it remains unclear how promotional activity on individual items may influence buyers' evaluations of bundles. This research effort examines the extent to which bundling can stimulate demand under specific promotion conditions (e.g., when the individual items are frequently promoted). We show that increased promotion activity on the individual items can significantly lower buyers' evaluations of bundle offers featuring the individual items. Managerial and research implications of these findings are discussed.

2 Magnitude and Frequency Effects

Buyers' perceptions of promotional activity on individual items can be influenced by (a) the expected magnitude of price variation, and (b) the expected frequency of price variation on individual items. The expected magnitude of price variation is likely to depend on how deeply the individual items are price-promoted. Similarly, the expected frequency of price variation is likely to depend on how often individual items go on sale. Previous research has shown that the magnitude and frequency of price promotion have a significant impact on buyers' evaluations of individual items (e.g., Alba et al., 1994; Kalwani and Yim, 1992; Krishna, Currim and Shoemaker, 1991; Krishna, 1991). To examine the role of these two promotional characteristics in the context of bundling, consider the following example. Two products (A and B) can be purchased separately for the regular prices \$x and \$y, respectively; bundle A+B can be purchased for \$z, where z represents a discount compared to \$x+y. Buyers' perceptions of savings offered on the bundle are likely to be based on a comparison of \$z relative to \$x+y (Yadav and Monroe, 1993). That is, \$x+y may serve as a reference for judging bundle savings. However, as discussed below, promotional activities on the bundle's individual items can significantly diminish the usefulness of \$x+y as a reference for judging bundle savings.

2.1 Magnitude Effect

First, in the above example, consider the situation where buyers expect large variations in the prices of items A and B when purchased individually. Under such conditions, buyers are less likely to regard \$x and \$y as representative of the items' regular prices (Biswas and Blair, 1991). Furthermore, the usefulness of \$x+y as a reference for judging bundle savings is also likely to be diminished. To account for the perceived price uncertainty, buyers may discount \$x+y (i.e., adjust downward) before using that amount as a reference for evaluating bundle savings (see, e.g., Biswas and Burton, 1993; Gupta and Cooper, 1992). The net effect of such adjustments will be lower transaction value (i.e., perceived savings; see Thaler, 1985) for the bundle, lower willingness to buy the bundle, and lower reservation price (i.e., the maximum price one is willing to pay) for the bundle. We refer to this as the magnitude effect.

H1: The larger the magnitude of expected price variation on the individual items, the lower (a) the buyers' transaction value for the bundle (b) the buyers' willingness to buy the bundle and (c) the buyers' reservation price for the bundle.

2.2 Frequency Effect

Continuing with the above example, the usefulness of \$x+y as a reference for judging bundle savings is also likely to be diminished when products A and B are frequently price-promoted. Considerable evidence indicates that the frequency of sales on individual items has a significant impact on the buyers' price expectations of these items (e.g., Alba et al., 1994; Biswas and Blair, 1991; Monroe, 1990; Krishna, 1991; Kalwani and Yim, 1992). Kalwani et al. (1990) have shown that consumers form expectations of a brand's price on the basis of its past prices and the frequency with which it is promoted. Their findings indicate that the higher the frequency of sales on individual items, the lower are the consumers' price expectations for the individual items. Consistent with these findings, Krishna (1991) found that the price consumers are willing to pay for a brand has a significantly negative correlation with consumers' perceived deal frequency of the brand. This implies that the higher the perceived deal frequency, the lower will be the consumers' willingness to pay for the items' promoted. Therefore, when individual items in a bundle are frequently promoted, the price consumers expect to pay for the bundle ($x+y) is likely to be lowered. Because $x+y is used as a reference to judge savings on the bundle, reductions in this reference should lower buyers' perceived transaction value for the bundle, buyers' willingness to buy the bundle, and buyers' reservation price for the bundle. We refer to this as the frequency effect.

H2: The higher the frequency of price variation on individual items, the lower (a) the buyers' perceived transaction value for the bundle, (b) the buyers' willingness to buy the bundle, and (c) the buyers' reservation price for the bundle.

3 Method

3.1 Subjects

To examine the above hypotheses, 205 undergraduate students at a major southwestern university in the United States participated in an experiment involving the evaluation of a two-item travel package (airfare and hotel accommodation, see Appendix). These items were selected on the basis of a pretest which showed that subjects were familiar with them, and considered them appropriate for bundling. After discarding 5 unusable questionnaires, a total of 200 responses was available for final analysis. The average age of subjects was 20.6 years and 55% were female.

3.2 Design and Procedure

A 2 (Frequency of airfare sales: frequently vs. rarely) x 2 (Frequency of hotel accommodations sales: frequently vs. rarely) x 2 (Magnitude of price variation on individual items: low vs. high) between-subjects design was used. Therefore, eight alternative bundling scenarios were created for studying the effects of these three factors on buyers' perceived transaction value for the bundle, their willingness to buy the bundle and buyers' reservation price for the bundle. Subjects were randomly assigned to one of the eight groups so that each contained 25 subjects. Based on pretests, a "low" magnitude of price variation on the individual items was manipulated by stating that prices ranged from \$350 ("lowest price observed") to \$450 ("highest price observed"). A "high" magnitude of price variation was represented by a \$200-\$600 price range. Frequency of sales was manipulated by indicating that the individual items went on sale either "frequently" or "rarely". The bundle price was kept constant at \$750 in all cells. A sample of the stimulus material is shown in the Appendix.

3.3 Manipulation Checks

We included manipulation check items in our questionnaire to see whether the three experimental factors were successfully manipulated. Using the guidelines provided by Perdue and Summers (1986), each manipulation check item was subjected to an analysis of variance. In each analysis, the effect size of the manipulated variable (ω^2) was much larger than the other main effect sizes; the interaction effects were almost zero. ω^2 is a statistic which ranges from zero (no treatment effect) to one (no error). It can be defined as the proportion of variance in the "dependent variable" (in this case, the particular manipulation check item being analyzed) accounted for by a given main or interaction effect. Ideally, for any given manipulation check item, we would like to find a sufficiently large ω^2 associated with the main effect of the manipulated factor corresponding to the manipulation check item being analyzed and a near-zero ω^2 for each of the other main effects. The results of our manipulation checks are summarized below (ω^2 associated with the manipulated variables are shown in bold). For conciseness, only main effects are reported.

Expected frequency of sales on airfare sales (**0.70**, 0.015, 0.000),
Expected frequency of sales on hotel accommodation sales (0.008, **0.672**, 0.027),
Expected magnitude of price variation on airfare sales (0.022, 0.026, **0.377**),
Expected magnitude of price variation on hotel accommodation sales (0.035, 0.027, **0.468**).

Furthermore, t-tests indicated that there were significant differences ($p < 0.0001$) between the levels of the manipulated factors. Based on these analyses of variance and t-tests, we concluded that the three factors were manipulated successfully in our study.

3.4 Dependent Measures

The three dependent variables in our study were (1) buyers' willingness to buy the bundle, (2) buyers' perceived transaction value for the bundle, and (3) buyers' reservation price for the bundle. Based on a pretest (n = 49) and on the literature (Dodds, Monroe and Grewal, 1991; Yadav and Monroe, 1993), we operationalized buyers' transaction value of the bundle and buyers' willingness to buy the bundle with three items for each construct. Bundle's transaction value was measured with the following items: (1) Overall, if you buy A and B as a package, you would be taking advantage of an attractive price reduction (strongly disagree, strongly agree); (2) Overall, if you buy A and B as a package, the deal you would be getting is (very poor, very good); (3) Overall, buying A and B as a package appears to be a good bargain (strongly disagree, strongly agree). Willingness to buy the bundle was measured with the following items: (1) The likelihood that you would purchase A and B as a package is (very low, very high); (2) Your intention to purchase A and B as a package is (very low, very high); (3) You would seriously consider buying A and B as a package (strongly disagree, strongly agree). Seven-point ratings scales were used for all items. Cronbach's alpha was 0.89 for transaction value and 0.94 for willingness to buy. Subjects' reservation prices for the bundle was measured by asking them to indicate the maximum price they would be willing to pay for the package (airline ticket plus hotel accommodation).

4 Results

4.1 Magnitude Effect

The results are reported in Tables 1, 2 and 3. Hypothesis 1 (a) states that the larger the magnitude of price variation on the individual items, the lower the buyers' perceived transaction value for the bundle. ANOVA results indicated that there was a significant main effect associated with the factor magnitude of price variation ($F_{1,199} = 5.264$, $p < 0.023$, $\eta^2 = 0.027$). No significant interaction effects were found. As the magnitude of price variation on the individual items increased from low ($350-$450) to high ($200-$600), the buyers' transaction value for the bundle decreased from 3.95 to 3.52 ($t = 2.24$, $p < 0.026$; see Table 1). Based on these results hypothesis 1 (a) is confirmed.

Table 1. Mean responses of buyers' transaction value of the bundle

Magnitude of Price Variation on Individual Items	Frequency of Price Variation on Individual Items				
	Consistent Scenario		Mixed Scenario		
	A Frequently B Frequently	A Rarely B Rarely	A Frequently B Rarely	A Rarely B Frequently	Row Means
$350-450	3.81 (1.31) [1]	4.21 (1.42) [3]	3.93 (1.23) [5]	3.82 (0.83) [7]	3.95
$200-600	2.68 (1.38) [2]	3.95 (1.51) [4]	3.84 (1.47) [6]	3.62 (1.15) [8]	3.52
Column Means	3.24	4.08	3.88	3.72	3.74

Products A and B refer to the airfare and hotel accommodation, respectively. Cell numbers are indicated in the last line of each cell (in parentheses). In each cell, the mean of the buyers' perceived transaction value of the bundle is reported with standard deviations shown in parentheses. The numbers of respondents in each cell is 25; N = 200.

Hypothesis 1 (b) posits that the higher the magnitude of price variation on the individual items, the lower the buyers' willingness to buy the bundle. The ANOVA results indicated a significant main effect associated with the factor magnitude of price variation ($F_{1,199} = 16.911$, $p < 0.0001$, $\eta^2 = 0.081$). The interaction effects were insignificant. As shown in Table 2, an increase in the magnitude of price variation on the individual items leads to a decrease in the buyers' willingness to buy the bundle (from 4.40 to 3.52, t = 3.82, $p < 0.0001$). These results support hypothesis 1 (b).

Table 2. Mean responses of buyers' willingness to buy the bundle

Magnitude of Price Variation on Individual Items	Frequency of Price Variation on Individual Items				Row Means
	Consistent Scenario		Mixed Scenario		
	A Frequently B Frequently	A Rarely B Rarely	A Frequently B Rarely	A Rarely B Frequently	
\$350-450	3.97 (1.71) [1]	5.28 (1.18) [3]	4.24 (1.71) [5]	4.12 (1.56) [7]	4.40
\$200-600	2.18 (1.26) [2]	4.13 (1.62) [4]	3.96 (1.59) [6]	3.79 (1.46) [8]	3.52
Column Means	3.08	4.71	4.10	3.95	3.96

Table 3. Mean responses of buyers' reservation prices for the bundle

Magnitude of Price Variation on Individual Items	Frequency of Price Variation on Individual Items				Row Means
	Consistent Scenario		Mixed Scenario		
	A Frequently B Frequently	A Rarely B Rarely	A Frequently B Rarely	A Rarely B Frequently	
\$350-450	702.6 (116.38) [1]	743.0 (107.41) [3]	718.0 (110.99) [5]	711.0 (109.23) [7]	718.65
\$200-600	632.8 (104.78) [2]	708.4 (111.35) [4]	682.0 (88.84) [6]	698.0 (126.23) [8]	680.3
Column Means	667.7	725.7	700.0	704.5	699.5

Hypothesis 1 (c) states that the higher the magnitude of price variation on the individual items, the lower the buyers' reservation price for the bundle. The ANOVA results indicated a significant main effect associated with the factor magnitude of price variation ($F_{1,199} = 6.094$, $p < 0.014$, $\eta^2 = 0.031$). No significant interaction effects were found. As shown in Table 3, an increase in the magnitude of price variation on the individual items leads to a decrease in the buyers' reservation price for the bundle from 718.65 to 680.30 (t = 2.45, $p < 0.015$). These results support hypothesis 1 (c).

4.2 Frequency Effect

Hypothesis 2 (a) posits that the higher the frequency of price variation on the individual items, the lower the buyers' perceived transaction value of the bundle. The ANOVA results showed that there was a significant main effect associated with the factor frequency of sales on hotel accommodations ($F_{1,199} = 7.246$, $p < 0.008$, $\eta^2 = 0.036$); the factor frequency of airfare sales was significant at $p < .10$ ($F_{1,199} = 3.329$, $p < 0.070$, $\eta^2 = 0.017$). The interaction effects were insignificant. In Table 1, note that the more frequently bundle items go on sale, the lower the buyers' perceived transaction value for the bundle tends to be. Further analysis using t-tests indicated that there are significant differences between column 1 (both items are frequently on sale) and column 2 (both items are rarely on sale) ($t = -2.86$, $p < 0.005$). These results support hypothesis 2 (a).

According to hypothesis 2 (b), the higher the frequency of price variation on the individual items, the lower the buyers' willingness to buy the bundle. The pattern of means reported in Table 2 is consistent with this hypothesis. Means of column 1 (3.08) and column 2 (4.71) were significantly different ($t = -4.97$, $p < 0.0001$). The ANOVA result showed significant main effects associated with the frequency of airfare sales ($F_{1,199} = 11.779$, $p < 0.001$, $\eta^2 = 0.058$) and the frequency of hotel accommodations sales ($F_{1,199} = 16.612$, $p < 0.0001$, $\eta^2 = 0.080$). All interaction effects were insignificant. These results support hypothesis 2 (b).

Hypothesis 2 (c) states that the higher the frequency of price variation on individual items, the lower the buyers' reservation price for the bundle. The pattern of means reported in Table 3 is consistent with this hypothesis. Means in column 1 ($667.7) and column 2 ($725.7) were significantly different ($t = -2.58$, $p < 0.011$). The ANOVA results indicated a significant main effect associated with the frequency of airfare sales ($F_{1,199} = 4.046$, $p < 0.046$, $\eta^2 = 0.021$); the factor frequency of hotel accommodations sales was significant at $p < .10$ ($F_{1,199} = 2.965$, $p < 0.087$, $\eta^2 = 0.031$). Thus, hypothesis 2 (c) is confirmed. A summary of the ANOVA results is provided in Table 4.

Table 4. Summary of ANOVA results

Sources of Variation	F[a]	Prob.	η^2 [b]
I. Transaction Value of the Bundle			
Frequency of Airfare Sales	3.329	0.070	0.017
Frequency of Hotel Accommodation Sales	7.246	0.008	0.036
Magnitude of Price Variation	5.264	0.023	0.027
II. Willingness to Buy the Bundle			
Frequency of Airfare Sales	11.779	0.001	0.058
Frequency of Hotel Accommodation Sales	16.911	0.0001	0.081
Magnitude of Price Variation	16.911	0.0001	0.081
III. Reservation Price for the Bundle			
Frequency of Airfare Sales	4.046	0.046	0.021
Frequency of Hotel Accommodation Sales	2.965	0.087	0.015
Magnitude of Price Variation	6.094	0.014	0.031

[a] F-value corresponding to Wilk's lambda with degrees of freedom (1, 199).

[b] Eta squared (η^2) is a statistic which ranges from zero (no treatment effect) to one (no error). It is used to provide information on the relative contribution of each factor to the variance in the dependent variable.

5 Conclusion

Economic analyses suggest that bundling strategies are generally more profitable than selling items individually (Adams and Yellen, 1976; Schmalensee, 1984). These suggestions are empirically supported by several studies which demonstrated that bundling strategies stimulate demand and increase profit (e.g., Simon, 1992; Venkatesh and Mahajan, 1993; Fuerderer, 1996; Wuebker, 1998). The results of this study suggest that under certain market conditions, the effectiveness of bundling strategies can be diminished considerably. As promotional activity increased on the individual items, buyers perceived less savings on the bundle, appeared less inclined to purchase the bundle, and were less willing to pay for the bundle. Therefore, it appears that buyers tend to buy the bundle when the price advantage of the single items over the bundle seems to be low and when it seems to be difficult to find these individual items on sale. Time savings, convenience and therefore lower transaction costs may be a reason for this behavior (Coase, 1960; Demsetz, 1968).

This research examined the effects of two components of promotional activity on the individual items: the magnitude and frequency of price variation. The results suggest that the buyers' evaluation of bundles may be influenced more strongly by the magnitude of price variation that buyers expect on the individual items. Future research efforts should be directed at further exploring these different effects associated with the magnitude and frequency of price variations on individual items. Managerially, the findings reported here highlight the potentially detrimental effects on bundling resulting from promotional activity on the individual items. Research efforts that seek to model the effectiveness of bundling strategies (e.g., Venkatesh and Mahajan, 1993) should also recognize the important connection between the outcomes of bundling strategies and the promotional context in which these strategies are implemented.

Appendix

Sample Stimulus Material

Assume that you wish to go from Austin to San Francisco for one week during the 1997 spring break. You want to fly to San Francisco and stay in a hotel for one week. Therefore, you have kept an eye on airline and hotel accommodation prices. You have gathered the following information:

Product	Lowest price observed	Highest price observed	Frequency of sales
A. Round trip airline ticket (Austin - San Francisco)	$ 350	$ 450	Rarely goes on sale
B. Hotel accommodation for one week (in San Francisco)	$ 200	$ 600	Frequently goes on sale

OR

PACKAGE DEAL AVAILABLE: Buy A and B as a package for $ 750

Manipulation Check Items

The observed price of product A seems to vary (very little, a lot),
The observed price of product B seems to vary (very little, a lot),
Finding product A on sale would appear to be (easy, difficult),
Finding product B on sale would appear to be (easy, difficult).

Seven-point ratings scales were used for all items.

References

Alba, J. W., S. M. Broniarczyk, T. A. Shimp and J. E. Urbany (1994). "The Influence of Prior Beliefs, Frequency Cues, and Magnitude Cues on Consumers' Perceptions of Comparative Price Data." Journal of Consumer Research, Vol. 21 (September), 219-235.

Adams, W. J. and J. L. Yellen (1976). "Commodity Bundling and the Burden of Monopoly." Quarterly Journal of Economics, Vol. 90 (August), 475-498.

Biswas, A. and E. A. Blair (1991). "Contextual Effects of Reference Prices in Retail Advertisements." Journal of Marketing, Vol. 55 (July), 1-12.

_____, and S. Burton (1993). "Consumer Perceptions of Price Claims: An Assessment of Claim Types Across Different Discount Levels." Journal of the Academy of Marketing Science, Vol. 21 (Summer), 217-230.

Burstein, Meyer Louis (1960). "The Economics of Tie-In Sales." Review of Economics and Statistics, Vol. 42 (February), 68-73.

Coase, R. H. (1960). "The Problem of Social Cost." The Journal of Law & Economics, Vol. 3 (October), 1-44.

Demsetz, H. (1968). "The Cost of Transacting." Quarterly Journal of Economics, Vol. 82 (February), 33-53.

Dodds, W. B., K. B. Monroe and D. Grewal (1991). "Effects of Price, Brand, and Store Information on Buyers' Product Evaluations." Journal of Marketing Research, Vol. 28 (August), 307-319.

Dolan, R. J. and H. Simon (1996). Power Pricing: How Managing Price Transforms the Bottom Line. New York, NY. The Free Press.

Drumwright, M. E (1992). "A Demonstration of Anomalies in Evaluations of Bundling."Marketing Letters, Vol. 3 (October), 311-321.

Fuerderer, R. (1996). "Option and Component Bundling under Demand Risk." Wiesbaden. Gabler.

Gaeth, G. J., I. P. Levin, G. Chakraborty and A. M. Levin (1990). "Consumer Evaluation of Multi-Product Bundles: An Information Integration Analysis." Marketing Letters, Vol. 2 (January), 47-57.

Guiltinan, J. P. (1987). "The Price Bundling of Services: A Normative Framework." Journal of Marketing, Vol. 51 (April), 74-85.

Gupta, S. and L. G. Cooper (1992). "The Discounting of Discounts and Promotion Thresholds." Journal of Consumer Research, Vol. 19 (December), 401-411.

Hanson, W. A. and R. K. Martin (1990). "Optimal Bundle Pricing." Management Science, Vol. 36 (February), 155-74.

Kalwani, M. U. and Chi Kin Yim (1992). "Consumer Price and Promotion Expectations: An Experimental Study." Journal of Marketing Research, Vol. 29 (February), 90-100.

________, Chi Kin Yim, H. J. Rinne and Y. Sugita (1990). "A Price Expectations Model of Customer Brand Choice." Journal of Marketing Research, Vol. 27 (August), 251-262.

Kenney, R. W. and B. Klein (1983). "The Economics of Block Booking." Journal of Law and Economics, Vol. 26 (October), 497-540.

Krishna, A. (1991). "Effect of Dealing Patterns on Consumer Perceptions of Deal Frequency and Willingness to Pay." Journal of Marketing Research, Vol. 28 (November), 441-451.

________, I. C. Currim and R. W. Shoemaker (1991). "Consumer Perceptions of Promotional Activity." Journal of Marketing, Vol. 55 (April), 4-16.

Monroe, K. B. (1990). Pricing: Making Profitable Decisions. Second Edition. New York, NY. McGraw-Hill.

Perdue, B. C. and J. O. Summers (1986). "Checking the Success of Manipulations in Marketing Experiments." Journal of Marketing Research, Vol. 23 (November), 317-326.

Schmalensee, R. (1984). "Gaussian Demand and Commodity Bundling." Journal of Business, Vol. 57 (January), 211-30.

Simon, H. (1992). "Preisbündelung." Zeitschrift für Betriebswirtschaft, Vol. 62 (November), 1213-1235.

Stigler, G. J. (1968). "A Note on Block Booking." The Organization of Industry. Homewood, IL: Richard D. Irwin, Inc.

Thaler, R. (1985). "Mental Accounting and Consumer Choice." Marketing Science, Vol. 4 (Summer), 199-214.

Venkatesh, R. and V. Mahajan (1993). "A Probabilisitic Approach to Pricing a Bundle of Products or Services." Journal of Marketing Research, Vol. 30 (November), 494-508.

Whinston, M. D. (1990). "Tying, Foreclosure, and Exclusion." American Economic Review, Vol. 80 (September), 837-859.

Wuebker, G. (1998). Preisbündelung: Formen, Theorie, Messung und Umsetzung. Wiesbaden. Gabler.

Yadav, M. S. and K. B. Monroe (1993). "How Buyers Perceive Savings in a Bundle Price: An Examination of a Bundle's Transaction Value." Journal of Marketing Research, Vol. 30 (August), 350-358.

Yadav, M. S. (1994). "How Buyers Evaluate Product Bundles: A Model of Anchoring and Adjustment." Journal of Consumer Research, Vol. 21 (September), 342-353.

Yadav, M. S. (1995). "Bundle Evaluation in Different Market Segments: The Effects of Discount Framing and Buyers' Preference Heterogeneity." Journal of the Academy of Marketing Science, Vol. 21 (3), 206-215.

How Buyers Evaluate Product Bundles: A Model of Anchoring and Adjustment

Manjit S. Yadav[1]

[1] **Manjit S. Yadav**, Department of Marketing, Texas A&M University, College Station, Texas. The research reported in this article is based, in part, on a doctoral dissertation that benefited from the contributions of K.B. Monroe. This paper was published in the Journal of Consumer Research, 1994, pp. 342-353.

1 A Model of Bundle Evaluation

1.1 Bundle Evaluation

How do buyers evaluate a bundle of items? the most common approach in earlier economic analyses (see, e.g., Adams and Yellen 1976; Schmalensee 1984) was to start with the additivity assumption – the overall utility of a bundle equals the sum of the bundle items individual utilities. The restrictiveness of this additivity assumption was later recognized and removes (Dansby and Conrad 1984; Guiltinan 1987; Hanson and Martin 1990). However, the question of how buyers´evaluation processes may account for additivity (or a lack thereof) has remained largely unexplored.

Goldberg et al. (1984), using hybrid conjoint analysis to investigate buyers´evaluation of a bundle of hotel amenities, concluded that "simple functions of respondents´self-explicated utilities are not good predictors of their preferences for the total bundle of hotel amenities" (p. S129). Gaeth et al. (1990) experimentally varied the attributes of items in a bundle to test the hypothesis that buyers average individual items evaluations to form a bundle´s overall evaluation. The evidence supported the averaging hypothesis and also suggested that the impact of an item on bundle evaluation may be greater than that expected on the basis of its monetary worth alone. However, we still do not know much about how buyers make such evaluations. Thus, the issue of how buyers evaluate a bundle of items must be explored in greater detail to understand how these evaluations may occur.

Table 1. A summary of judgment tasks and methodological approaches used to assess the anchoring and adjustment hypothesis.

		Basis for evidence for hypothesis testing		
Empirical investigation	Judgement task	ANOVA	Regression	Protocol analysis
Biswas and Burton (1993)	perception of price claims	√		
Block and Harper (1991)	estimating uncertain quantities	√		
Switzer and Sniezek (1991)	motivation to complete task	√		
Mano (1990)	establishment of goals	√		√[a]
Hogarth and Einhorn (1989)	belief updating	√		
Wright and Anderson (1989)	probability assessment	√		
Johnson and Schkade (1988)	utility assessment	√		√[b]
Lopes (1987)	probability assessment	√		
... and Neale (1987)	price estimation	√		
..., Hoch, and Ragsdale (1986)	preference for new products	√	√	
Einhorn and Hogarth (1985)	belief change	√	√	√[a]
Lopes (1985)	probability assessment	√		
Friedlander and Stockman (1983)	clinical Judgement	√		
Joyce and Biddle (1981)	auditing	√		
Lopes and Ekberg (1980)	risky decision making	√		
Tversky and Kahneman (1974)	estimating uncertain quantities	√		f
This study	bundle evaluation	√	√	√

[a] Protocols were not coded; only excerpts were presented.

[b] Protocols components were not coded; instead, complete protocols were categorized into two broad groups: heuristic and nonheuristic.

1.2 Anchoring and Adjustment

Even when bundles have only a few items, the amount of information available for processing can be substantial, and it is conceivable that buyers may look always to simplify the evaluation task. If buyers wish to incorporate all the available information into their evaluation of the bundle, (e.g., when making a big-ticket purchase), the information-processing task can be rather daunting – unless a simplifying heuristic enables the buyer to approach the evaluation task as a series of smaller (and simpler) evaluations. The anchoring and adjustment heuristic (Tversky and Kahneman 1974) may enable buyers to accomplish such an evaluation task.

As shown in Table 1, anchoring and adjustment, which involves an initial assessment followed by one or more adjustments, has proved applicable in a variety of judgment tasks. Indeed, Johnson and Puto (1987) have argued that anchoring and adjustment could "underlie many of the structural models as well as processing strategies reported by decision researchers" (p. 255) In fact, the variety of averaging, additive, and multiplicative models typically encountered in Anderson´s (1981) information integration paradigm can be accounted for by proposing that the underlying processing strategy is anchoring and adjustment (Lopes 1982).

Lopes′(1982) model of an anchoring and adjustment process can be readily adapted to the context of bundle evaluation. As shown in Figure 1, three stages can be identified: (1) scanning, (2) anchor selection, and (3) anchoring and adjusting.

In the scanning stage, buyers determine which items are contained in the bundle, but evaluations are generally not made. During the anchor selection stage, buyers initiate the evaluation process by selecting one bundle item perceived to be most important for the evaluation task. Finally, during the anchoring and adjusting stage, evaluation of this selected item serves as a starting point (i.e., as an anchor evaluation) to initiate the evaluation of the bundle the remaining bundle items are subsequently evaluated in decreasing order of their perceived importance, and upward or downward adjustments are made to reflect the new information encountered (see Fig. 1). This iterative nature of anchoring and adjustment enables buyers to sequentially evaluate and integrate many pieces of information while placing few demands on short-term memory (Lopes 1982). Therefore, we hypothesize that

> H1: Buyers will form an overall evaluation of a set of bundle items by (a) examining the items in decreasing order of their perceived importance and (b) adjusting their bundle evaluations in the direction of the succeeding item evaluations.

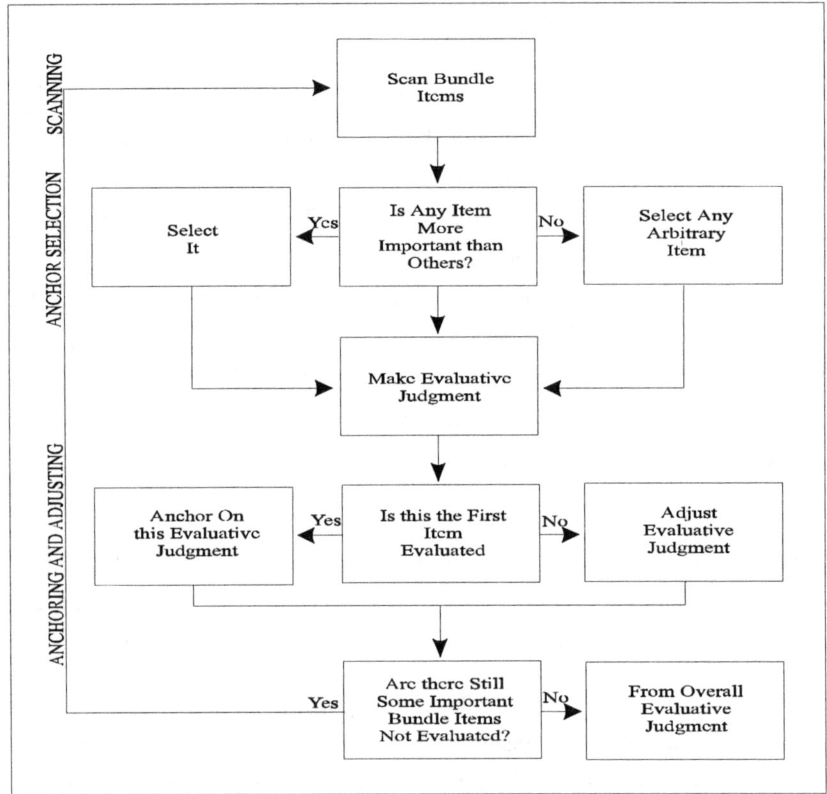

Fig. 1. Anchoring and adjustment model of bundle evaluation

1.3 Weighted-Average Representation

To operationalize anchoring and adjustment, Hogarth and Einhorn (1989) noted that it is necessary to determine (1) how the sequential evaluations are encoded prior to their integration with the anchor evaluation and (2) how the adjustments actually occur during the integration stage. On the basis of this approach, these authors showed that final judgments can be expressed algebraically in a weighted-average form. Einhorn and Hogarth (1985) observed that "anchoring and adjustment processes often result in outcomes that can be modeled as weighted averages" (p. 437). Further, Johnson and Puto (1987) have argued that averaging "is very consistent with anchoring and adjustment" (p. 256). Thus, averaging models and anchoring and adjustment both capture compensatory processes in different (but consistent and complementary) ways.

In the anchoring and adjustment process, it is assumed that adjustments are made after each new piece of evidence is evaluated. as this processing strategy enables decision makers to decompose a relatively complex task into several smaller and simpler components, it is likely to be used when many pieces of evidence have to be integrated. However, for simpler tasks, decision makers may anchor on the most important piece of evidence and then make only one adjustment on the basis of an ggregate evaluation of the remaining pieces of evidence. Hogarth and Einhorn (1989) showed that this one-adjustment form of anchoring and adjustment also yields a weighted-average algebraic form. Hence,

> H2: Adjustments made while evaluating a bundle of items will result in weighted averaging; that is, the overall evaluation of a bundle′s items will be a weighted average of the individual items′evaluation.

1.4 Bias Caused by Insufficient Adjustments

One finding frequently encountered in applications of anchoring and adjustment is that overall evaluations appear biased toward the anchor evaluation. Tversky and Kahneman (1974) have reasoned that subjects may first anchor on the first few items and then adjust for the remaining items to be judged. As adjustments tend to be insufficient, the final evaluation is likely to be biased in the direction of the initial anchor estimate.

Similarly, Davis et al. (1986) found support for their hypothesis that individuals′predictions about their spouses′reactions to new product concepts are made by anchoring on their own preferences and then making (insufficient) adjustments on the basis of other′factors. In another study, experienced realtors anchored on the listing price of residential properties and then adjusted these estimates on the basis of other information to assess the dollar value of the various properties (Northcraft and Neale 1987). Because these adjustments were insufficient, final dollar estimates we biased in the direction of the listing price. Finally, Joyce and Biddle (1981) provided two groups of experienced auditors with different anchor estimates and then asked them to estimate the incidence of executive-level management fraud. These estimates tended to be biased in the direction of the provided anchor information. Evidence consistent with the notion of insufficient adjustments ... comes from studies conducted in other substantive contexts, such as clinical judgments (Friedlander and Stockman 1983) and judgments under risk (Carlson 1990). Therefore,

> H3: Adjustments made while evaluating a bundle of items will be insufficient, in that the overall bundle evaluation will be biased in the direction of the item evaluated first.

Two experiments were conducted to test these hypotheses focusing on the anchoring and adjustment process. In designing these experiments, an effort was made to incorporate the different methodological approaches used in previous empirical investigations of anchoring and adjustment (ANOVA, regression, and protocol analyses; see Table 1). Specifically, the first experiment explored the sequential nature and algebraic representation of the hypothesized anchoring and adjurment process (Hypotheses 1 and 2): The second experiment investigated the influence of insufficient adjustments (Hypothesis 3).

2 Study 1

2.1 Design

By means of the Ci2 software package (Sawtooth Software 1987), a 3 (bundle context) x 2 (anchor context) between-subjects design was employed to examine how buyers evaluate a bundle of items. To illustrate the nature of the manipulated factors, consider a hypothetical bundle with three items, A, B , and C, such that A is perceived as most important in the bundle-evaluation task, followed in importance by items B and C, respectively. That is, item A is the a priori anchor. On the basis of the pretests described below, a computer bundle consisting of a computer (A), printer (B), and printer stand (C) and a furniture bundle consisting of a bed (A), chest (B), and nightstand (C) were chosen as products for this experiment.

The three levels of the factor bundle context represent three types of evaluation situations: (1) evaluation of individual items A, B, and C (i.e., no bundle evaluation), (2) evaluation of the bundle of items A and B, and (3) evaluation of the bundle of items A, B, and C. The two levels of the factor anchor context ("excellent" and "poor") represent siuations whre the anchor (i.e., item A) was manipulated to be either excellent or poor. Descriptions of the add-on items (B and C) were controlled to suggest moderate quality.

2.2 Bundle Selection and Development of Product Descriptions

Four pretests were conducted to select appropriate bundle products, to test descriptions of these products, and to develop measures for the variables of interest. Business undergraduate students at a state university served as subjects in these pretests. On the basis of the first pretest (n=70), a computer bundle (computer, printer, and printer stand) and a furniture bundle (bed, chest, and nightstand) were selected for use in the experiment. This pretest showed that subjects were familiar with the items, the items were perceived as differentially important for evaluating a given bundle, and there was a clearly defined anchor in each bundle (computer and bed).

For the second pretest (n=27), we developed paragraph-length descriptions (approximately 100 words) of the poor and excellent anchors (i.e., computer and bed). No price or brand information was presented. Similar descriptions for the add-on products were also developed to suggest "moderate" quality. Subjects were instructed to read each description and then evaluate the described product on three seven-point ratings scales (durability, quality, dependability/ workmanship), which yielded satisfactory internal consistency (Cronbach's $\alpha > 0.8$). The product descriptions were subsequently modified for a third pretest (n=9) was conducted to ensure that the software operated properly and that instructions provides to the subjects were adequate to guide them through their experimental task.

2.3 Measures

On the basis of the pretests described above, multi-item measures were developed for the following variables:

1. Overall bundle evaluation: buyers' overall assessment of excellence of all bundle items considered together as a set.
2. Item evaluation: buyers' overall assessment of excellence of an individual item in the bundle.
3. Relative item importance: buyers' perception of how important a bundle item is relative to other bundle items for evaluating the overall bundle.
4. Order of examination: the order in which bundle items are examined during the evaluation process.

Three ratings scales focusing on quality, durability, and dependability/ workmanship were used to operationalize overall bundle evaluation and item evaluation. Relative item importance was measured by means of three different approaches: (1) subjects rank ordered bundle items in terms of their perceived relative importance in the bundle, (2) subjects allocated 100 points between the bundle items to reflect their perceived relative importance in the bundle, and (3) subjects estimated each item′s average retail price. As described below, the order of examination was recorded by the software.

2.4 Procedure

The computer-administered experimental task began with two warm-up exercises designed to familiarize subjects with the keyboard′s operation. Subjects then examined and evaluated various items on the basis of short descriptions displayed on their computer screens. No price or brand information was presented. As described below, tasks in the six cells of the 3 (bundle context) x 2 (anchor context) design ranged from the evaluation of individual items to the evaluation of bundles with three items.

Individual-Item Conditions. Subjects examined and evaluated short descriptions of a computer, printer, printer stand, bed chest, and nightstand one item at a time. The order of presentation varied for subjects. Because subjects did not evaluate bundles, item evaluations ere the only measures in these cells.

Two-Item Bundle Conditions. Subjects examined and evaluated two bundles: a computer bundle (computer and printer) and a furniture bundle (bed and chest), one bundle at a time. The software first listed on the computer screen (but did not describe) the items in a bunsle, randomizing the order in which the items were listed: Subjects were then permitted to examine descriptions of the bundle items in any order of their choice, one item at a time. After examining all items, subjects first evaluated the overall bundle and then the individual items. Finally, measures of relative item importance were obtained.

Three-Item Bundle Conditions. The procedure was similar to the two-item bundle case except that bundles in these conditions had three items.

One hundred seventy business undergraduate students at a state university participated in this experiment with a chance to win $100 in a lottery offered as an incentive to participate. After discarding incomplete or unusable responses, 153 subjects′ responses were available for the final analyses. With cell sizes ranging from 22 to 30, procedures appropriate for unbalanced designs were used in all subsequent analyses reported below. On average, subjects complete their assigned experimental tasks in about 18 minutes.

2.5 Results

2.5.1 Reliability and Manipulation Check

Reliability of the variables overall bundle evaluation and item evaluation was satisfactory (Cronbach′s $\alpha > 0.85$). Manipulation checks revealed that the mean evaluation of the computer was 2.08 in the three cells with a poor anchor item and 6.18 in the other three cells featuring an excellent anchor item ($t(151) = 34.88$, $p < .001$). Corresponding figures for the bed were 1.77 and 6.57 in the poor and excellent conditions, respectively ($t(151) = 51.90$, $p < .001$). Mean ratings of the add-on items′ moderate descriptions were as follows: printer (4.15), printer stand (3.78), chest (3.30), and nightstand (3.80).

The three measures of relative item importance were consistent with the pretest results where each bundle had a clearly defined anchor (computer and bed). The first measure, based on ranks assigned to the various bundle items, revealed differences in perceived importance. A majority of the subjects regarded the computer and the bed as the items most important in their respective bundles (two-item bundles: computer, 84 percent; bed, 96 percent). The printer and chest ranked second, and the printer stand and nightstand ranked third. The second measure, based on point allocation, indicated that the computer was perceived as the major item in the computer bundle, followed by the printer and printer stand. In the furniture bundle, the bed was perceived as most important, followed by the chest and nightstand. Subjects assigned more than half of the total points in each bundle to the anchor items. Finally, results based on the price-based estimates mirrored those of the point allocation measure.

2.5.2 Order of Examination

The order in which subjects examined the bundle items is shown in Table 2. While evaluating the bundles with two items, a majority of the subjects examined the anchor item (computer or bed) first, followed by either the printer or the chest (see Table 2, part A). Kendall's W, computed with the order of examination as the dependent variable, was 58 ($\chi_1^2 = 26.27, p < .001$) for the computer bundle items and .40 ($\chi_1^2 = 17.82, p < .001$) for the furniture bundle items. This evidence suggests that there was agreement across subjects regarding the sequence in which bundle items were examined-items perceived as more important were examined before the less important items. Spearman's rank-order correlation between relative item importance (based on rank) and order of examination was .58 ($p < .001$) for the computer bundle items and was .67 ($p < .001$) for items in the furniture bundle.

The pattern of results for three-item bundles (see Table 2, part B) was similar. Kendall's W, with order of examination as the dependent variable, was .52 ($\chi_2^2 = 42.49, p < .001$) for the computer bundle and, again, was .52 ($\chi_2^2 = 47.52, p < .001$) for the furniture bundle. Spearman's rank-order correlation between relative item importance and order of examination was .75 ($p < .001$) for both the computer bundle and the furniture bundle.

Thus, subjects examined the most important bundle item first, followed by other bundle items in decreasing order for perceived relative importance (Hypothesis 1a). this order of examination is consistent with the anchoring and adjustment process.

Table 2. Order of examination of bundle items

Bundle Items		Order of examination[a] (% of subjects) 1	2	3
A.	Bundles with two items:			
	Computer	88.6	11.4	
	Printer	11.4	88.6	
	Bed	81.8	18.2	
	Chest	18.2	81.8	
B.	Bundles with three items:			
	Computer	82.9	7.3	9.8
	Printer	7.3	82.9	9.8
	Printer stand	9.8	9.8	80.4
	Bed	84.8	8.7	6.5
	Chest	6.5	71.8	21.7
	Nightstand	8.7	19.5	71.8

[a] The order of examination was recorded by the software.

2.5.3 The Nature of Upward and Downward Adjustments

Overall Bundle Evaluation. The results displayed in Table 3 illustrate the upward and downward adjustments. In the first column of numbers, which represents the situation where moderate items were combined with an excellent anchor, downward adjustments are evident in both the computer and furniture bundles. The F-test for differences and the linear contrasts tests for the decreasing trend of overall bundle evaluations were significant. The second column in Table 3 illustrates the upward adjustments when the same moderate items were combined with a poor anchor. As before, the F-tests, as well as the linear contrasts, were significant.

Table 3. Effect of moderate item(s) on overall bundle evaluation when anchor is excellent or poor

Bundle context		Overall bundle Evaluation[a] Excellent anchor		Poor Anchor	
A.	Computer	6.05	(.71)	2.13	(.88)
	Computer and printer	5.07	(.70)	2.75	(.98)
	Computer, printer and printer stand	4.85	(.66)	2.63	(.94)
	F = test of part A:				
	F	22.96[b]		3.47[c]	
	P	.00		.04	
	η	.62		.29	
	t = test of part A:d				
	T	6.26[e]		2.01[f]	
	P	.00		.03	
	R	.35		.23	
B.	Bed	6.65	(.47)	1.79	(.64)
	Bed and chest	4.79	(.76)	2.49	(.73)
	Bed, chest and nightstand	4.29	(.69)	2.42	(.72)
	F = test of part B:				
	F	104.14[b]		8.52[c]	
	P	.00		.00	
	η	.86		.43	
	t = test of part B:d				
	T	13.45[d]		3.36[e]	
	P	.00		.00	
	R	.85		.36	

Note. – Standard deviations are shown in parentheses; η and r refer to effect sizes appropriate for F- and t-tests, respectively. The effect sizes, when squared, indicate the proportion of variance explained.

[a] Overall bundle evaluation was operationalized as the average of three seven-point ratings scales focusing on subjects´ perceptions of quality, durability, and dependability/ workmanship.

[b] df = 2,72

[c] df = 2,75

[d] Linear contrast with weights 1, 0, and –1 was used to test the declining trend in the excellent anchor condition. The corresponding weights were –1, 0, and 1 in tests of the increasing trend in the poor anchor condition.

[e] df = 72.

[f] df = 75.

Thus, at the aggregate level, the bundle evaluations were consistent with the hypothesized upward and downward adjustments (Hypothesis 1b).

To further explore the nature of adjustments, analyses at the level of each subject were also conducted. In the four cells where bundle evaluations were made, we compared each subject's anchor evaluation with that subject's overall bundle evaluation. In general, subjects′ evaluations were consistent with the aggregate results reported above, but the degree of consistency varied with the type of anchor (excellent or poor). Specifically, subjects were more likely to adjust their evaluations downward with the addition of moderate items (two-item bundles: 100 percent). However, a smaller percentage of subjects adjusted upward with the addition of moderate items to the poor anchors (two-item bundles: 79 percent; three-item bundles: 82 percent).

In general, it can be observed that evaluations of the two- and three-item bundles tended to be quite similar (see Table 3). Although the third item was the least important, its diminished impact may also have reflected subjects′ tendency not to change the bundle evaluation after the first two items had been evaluated. In the poor anchor condition, the third item resulted in a slight decrease in bundle evaluation, although the change was not statistically significant ($p > 0.5$) – which perhaps reflected subjects′ reluctance to change the bundle evaluation. Lopes (1982) alluded to a similar effect when she noted "that there is an increasing tendency during judgement for the impression being constructed to crystallize and become resistant to change" (p.31). Thus, when upward adjustments are needed, subjects may be more likely to focus primarily on the first few items in the bundle.

Effect of Bundle Context and Anchor Context. As expected from the hypothesized upward and downward adjustments, an ANOVA yielded a significant interaction between the bundle context and anchor context (computer bundle: $F(2,147)=18.52, p<.001, \eta^2=.20$; furniture bundle: $F(2,147)=81.15$, $p<.001$, $\eta^2=.52$). Overall bundle evaluation was the dependent variable, with the exception of two cells where subjects evaluated only individual items. In these cells, anchor evaluation was used instead as the dependent variable. The significant interaction suggests that the effect of moderate items on bundle evaluations depended on whether the anchor item was excellent or poor. In the excellent conditions, evaluations of the moderate items resulted in an increase in the evaluation of the overall bundle. This finding provides support for Hypothesis 1b.

To explore whether the pattern of results varied across the two bundles, we conducted a 2 (product type) $\times 3$ (bundle context) $\times 2$ (anchor context) ANOVA. As expected, the results were dominated by the strong effects associated with anchor context ($F(1,294) = 1142.69, p < .001, \eta^2 = .79$), bundle context ($F(2,294) = 18.11, p < .001, \eta^2 = .11$), and the interaction between them ($F(2,294) = 81.05, p < .001, \eta^2 = .36$). The results also revealed small but significant effects associated with product type ($F(1,294) = 4.16, p < .04, \eta^2 = .014$) and an interaction between product type and bundle context ($F(2,294) = 3.63, p < .03, \eta^2 = .024$). Finally, the three-way interaction suggested that the joint influence of anchor context and bundle context varied with the type of product ($F(2,294) = 5.43, p < .005, \eta^2 = .04$).

2.5.4 Weighted-Average Representation

By means of multiple regression, overall bundle evaluation was modeled as a dependent variable with evaluations of the individual items as the predictor variables. Two sets of regression analyses were conducted to test Hypothesis 2, that a bundle's overall evaluation is a weighted average of its items′ individual evaluations. In Table 4, regression coefficients in the first column pertain to the first analysis in which no constraints were imposed on the regression models. Interaction effects were not significant ($p < .25$). The coefficients suggest that the computer was the dominant item in the computer bundles. A comparison of the bed and chest coefficients indicates that both items were weighted equally ($p < .05$) – which is a surprising result given that most subjects examined the bed first (see Table 2).

Another point to note is that the evaluation of the three-item bundles appears to be driven mostly by the first two items. This evidence is consistent with our earlier observation that subjects may avoid making adjustments after the first two items. Larger bundles perhaps trigger noncompensatory evaluation in which subjects take into consideration only a part of the total information.

Table 4. Regression coefficients of the full and reduced models

Bundle Items		Regression coefficients[a] Full model	Reduced model	F* to compare the full and reduced models[b]
A.	Bundles with two items:			
	Computer	.55	.56	
	Printer	.37	.44	
	R^2	.77	.70	F*(1,42)= .470, F*<F
	Bed	.50	.50	
	Chest	.50	.50	
	R^2	.86	.87	F*(1,44)= .001, F*<F
B.	Bundles with three items:			
	Computer	.51	7.3	
	Printer	.33	82.9	
	Printer stand	.10	9.8	
	R^2	.74		F(1,44)= .560, F*<F
	Bed	.39	.39	
	Chest	.42	.45	
	Nightstand	.14	.16	
	R^2	.83	.91	F*(1,44)= .200, F*<F

[a] All regression coefficients were significant at p<.05 (except the nightstand [p<.08] and the printer stand [p<.22]); in the reduced model, the sum of the coefficients was constrained equal 1.

[b] $F^* = [(SSE_R - SSE_F)/(df_m - df_F)]/(SSE_F/df_F)$, where SSE denotes sums of squares error and df denotes degrees of freedom; subscript "F" refers to the full model and "R" to the reduced model /Neter, Wasserman, and Kutner 1990). F+<F implies that the full and reduced models are statistically compareable (p<.05).

The second column in Table 4 pertains to the "reduced model" in which the regression coefficients were constraint to equal 1 to represent a weighted-average algebraic form (see Einhorn and Hogarth 1985; Hogarth and Einhorn 1989; Lopes 1982). The statistic F* (Neter, Wassermann, and Kutner 1990; see Table 4) provides a statistical comparison of these two regression models and a test of the weighted-average hypothesis (Hypothesis 2). Specifically, if F+ is not statistically significant, the weighted-average model is an appropriate algebraic representation of the data. Comparison of the F+-values with corresponding values of F supported the weighted-average hypothesis for all the two-item and three-item bundles (see Table 4).

There was also evidence that a simple-average model (i.e., with all weights equal) may often represent the data adequately (Einhorn and Hogarth 1975; Hoffman 1960). For instance, in the case of the two-item bundles, the simple-average form was statistically comparable to the weighted-average form (computer bundle: $F(1,42)=.001, NS$). However, this was not hte case when the bundle had three items (computer bundle: $F(2,44)=8.95, p<.001$; bedroom bundle: $F(2,44)=3.40, p<0.5$).

2.5.5 Analysis of Concurrent Protocols

We instructed a separate group of 10 subjects to think aloud while evaluating the computer and furniture bundles. The resulting concurrent protocols were tape-recorded and transcribed. Following the procedures recommended for protocol analysis (Ericsson and Simon 1984), we divided subjects′ transcripts into separate task-specific statements. This procedure yielded 71 process statements. Focusing on statements related to bundle evaluation, we created the following four coding categories: (a) evaluation of the anchor item or another individual item; (b) comparison of an item's evaluation relative to a reference evaluation (i.e., an anchor item's or bundle's current evaluation); (c) adjustment of bundle evaluation; and (d) process statement related to bundle evaluation but not a, b, or c.

Two researchers, working independently, assigned the 71 process statements to the four coding categories. There was agreement regarding the assignment of 62 of the 71 statements (87 percent). Coding for the remaining nine statements was resolved through discussion. Statements related to the adjustment of bundle evaluation (category c) occurred most frequently, accounting for almost half of all the statements (33; 47 percent). Comparative evaluation of an item relative to a reference evaluation was the second most common type of statement (23; 32 percent). Evaluation of an anchor item or another individual item by itself (i.e., without specifically mentioning a point of refernece; coding category a) occured in 10 (14 percent) of the process statements. Five statements (7 percent) were coded as belonging to category d.

Overall, the protocol analysis provided several additional insights into the hypothesized anchoring and adjustment process. First, the analysis indicates that adjustments may occur often during bundle evaluation; almost half of all the statements reflected an adjustment of bundle evaluation. This evidence is consistent with the hypothesis (1b) that subjects approach the task of bundle evaluation as a series of adjustments. Second, the protocol data permitted the examination of the type of encoding implied by Hogarth and Einhorn's (1989) operationalization of anchoring and adjustment – the evaluation of an item relative to a reference evaluation. About one-third of all process statements reflected this type of a comparative evaluation, which is suggestive of its frequent occurrence during bundle evaluation.

2.6 Summary

Subjects showed selectivity in examining the various bundle items – items perceived as more important were examined prior to the less important items – which is consistent with Hypothesis 1a. The aggregate and individual-level results, along with the protocol data, suggest that adjustments occur frequently during bundle evaluation (Hypothesis 1b). Depending on how the anchor evaluation compared with the evaluations of other items, these adjustments were either upward or downward. Furthermore, the anchor evaluation appeared to affect subject's tendency to adjust. When the anchor evaluation was better than the evaluation of other items, almost all subjects adjusted downward. However, when the add-on items were better than the anchor, some subjects did not adjust upward. Finally, the regression analysis showed that overall bundle evaluations can be expressed algebraically as aweighted average of the individual items' evaluations (Hypothesis 2).

3 Study 2

An important limitation of study 1 was its inability to fully explore the nature of adjustments that occur during anchoring and adjustment. As discussed earlier, overall evaluations formed through an anchoring and adjustment process are typically biased toward the evaluations made first, because adjustments biased on subsequent evaluations tend to be insufficient (Tversky and Kahneman 1974). On the basis of this notion of insufficient adjustments, we developed Hypothesis 3, which states that the overall evaluation of a bundle is biased in the direction of items evaluated early in the evaluation process. However, in study 1, items order of examination was confounded with their importance, in that items perceived as being more important were also examined earlier. Hence, the biasing effect (if any) associated with the order of examination alone could not be examined in the first investigation. Given that insufficient adjustments constitute a key characteristic of anchoring and adjustment, we conducted a second study to focus specifically on this issue.

3.1 Method and Procedure

Given the methodological similarities of this study with the first, we will highlight only the distinguishing features. The design, product stimuli, and procedure employed in this study remained largely unchanged – but with two important differences. First, two cells where subjects evaluated only individual items were dropped, which yielded a 2 (bundle context) × 2 (anchor context) design. That is, only two-item and three-item bundle evaluations were made. The second difference pertains to the order in which subjects acquired information about the individual items. In the first study, subjects were free to determine the order in which the items were evaluated, and the anchor item (computer, bed) was usually both the most important item and the item evaluated first. In the second study, however, subjects were shown bundle item's descriptions one at a time in a randomized order determined by the software. Subjects were subsequently asked to evaluate the overall bundle and the individual items.

With the order randomized in this manner, the effect of order on overall bundle evaluation and the notion of insufficient adjustments could be explored. For example, consider the case of a bundle featuring an excellent computer and moderate add-on items. Depending on which item was presented first, subjects' initial evaluations could vary considerably (from moderate to excellent). Hence, if insufficient adjustments (from the initial evaluation) result in a bias, examining the excellent anchor earlier (later) should result in higher (lower) overall evaluations. In contrast, when the anchor is poor, examining the anchor earlier (later) should result in lower (higher) overall evaluations.

One hundred forty business undergraduate students served as subjects, of which 127 provided complete and usable responses. A change to win $100 in a lottery was offered as an incentive to participate. Subjects from the first study were not recruited for this follow-up investigation.

3.2 Results

Insufficient Adjustments. The results are reported in Table 5. In the case of the two-item bundles in the excellent anchor condition, bundle evaluations were higher when the anchor item was evaluated first. The magnitudes of the effect size *r* (.17 and .20) suggest a weak-to-moderate effect, but the means were not statistically different. In the case of the three-item bundles, examining the excellent anchor item earlier also resulted in higher evaluations with linear contrasts. All F-test and then tested declining overall evaluations with linear contrasts. All F-tests and contrasts were statistically significant. We therefore conclude that, in the case of the three-item bundles with an excellent anchor, examining the anchor earlier resulted in higher overall bundle evaluations and examining later resulted in lower overall bundle evaluations. Hence, Hypothesis 3 was supported for the three-item bundles but not for the two-item bundles.

The prediction for the poor anchor condition was that examining the anchor earlier would result in a lower overall evaluation and examining it later would result in a higher overall evaluation. As can be observed in Table 5 (part A), evaluations of the two-item bundles remained statistically unchanged by the order of examination. However, in the case of the three-item bundles, examining the anchor item earlier generally led to lower bundle evaluations. The F-test suggested significant differences between the means, and the linear contrasts supported an increasing trend. We conclude that, as in the case of the excellent anchor condition, the three-item bundle evaluations were not. Later, we discuss why the biasing effect, which is due to insufficient adjustments, may be more likely to occur in the case of the three-item bundles.

Order of Evaluation and Perceived Importance. We also explored how the order of evaluation and perceived importance of an item jointly influence that item's impact on overall bundle evaluation. For example, it is reasonable to expect that the most important item (the anchor) will have the greatest impact on the overall evaluation of a bundle of items. But, if the order of evaluation is also important in determining how much impact an item has, the anchor's impact may diminish when it is not examined first.

We ran separate regression analyses for the two- and three-item bundles with two different sets of subjects: those who examined the anchor item first and those who examined the anchor item last. In the case of the two-item bundles, the computer's regression coefficient decreased in magnitude when the anchor was examined last (.66-.59); the bed's coefficient showed a similar decline (.64-.56). This declining trend was more pronounced in the case of the three-item bundles (computer, .58-.42; bed, .38-.14).

4 Discussion

4.1 Anchoring and Adjustment

Results of this investigation provided insights about the anchoring and adjustment heuristic in the context of bundle evaluation. Subjects examined items in decreasing order of perceived importance, making insufficient upward or downward adjustments to form the overall bundle evaluation. When faced with an excellent anchor and moderate add-on items, subjects readily adjusted the overall bundle evaluation downward. However, the tendency to adjust upward was considerably less when the anchor was poor and the ad-on items were moderate. Furthermore, the detrimental effect of moderate items on excellent anchors was more pronounced than their enhancing effect on poor anchors. Finally, by varying the order of presentation of the bundle items, it was shown that the biasing effects resulting from insufficient adjustments were stronger for three-item bundles than for two-item bundles.

If moderate items are perceived as "losses" when combined with excellent anchors and as "gains" when combined with poor anchors, these results are compatibles with the notion that the impact of perceived losses is greater than the impact of perceived gains (Kahneman and Tversky 1979). These results are also consistent with the more pronounced effects of negative information in the integration of information paradigm (Anderson 1981). Gaeth et al. (1990) report a similar effect in their study of bundle evaluation. Hence, when firms seek out possible items for bundling, they should recognize that it is easier to "hurt" an anchor than to "help" it. Providing consistent levels of quality in a bundle is therefore important.

Table 5. Effect of insufficient adjustments on overall bundle evaluation

Order of examination		Overall bundle evaluation[a]			
		Excellent anchor		Poor anchor	
		Computer bundle	Furniture bundle	Computer bundle	Furniture bundle
A.	Two item bundles:				
	Computer/ bed examined first	5.28 (.70)	5.17 (.96)	2.62 (.90)	2.45 (.66)
	Computer/ bed examined last	5.03 (.47)	4.85 (.89)	2.64 (.82)	2.31 (.73)
	t-test for part A:				
	t_{30}	1.09	.96	.07	.54
	p	.14	.17	.47	.29
	r	.20	.17	.01	.10
B.	Three-item bundles:				
	Computer/ bed examined first	5.23 (.37)	4.44 (.84)	2.67 (.75)	2.32 (.52)
	Computer/ bed examined second	4.76 (.77)	3.95 (.92)	3.67 (.69)	2.73 (.52)
	Computer/ bed examined last	4.28 (.90)	3.33 (.62)	3.20 (.87)	2.83 (.42)
	F-test for part B:				
	F	4.05[b]	3.70[b]	4,12[c]	3.43[c]
	p	.03	.04	.03	.05
	η	.47	.46	.47	.44
	t-test for part B:[d]				
	t	2.84[e]	2.72[e]	1.62[f]	2.50[f]
	p	.00	.00	.06	.01
	r	.47	.46	.29	.42

Note. – Standard deviations are shown in parentheses; η and r refer to effect sizes appropriate for F- and t-tests, respectively. The effect sizes, when squared, indicate the proportion of variance explained.

[a] For the two-item bundles in the excellent anchor condition, 19 and 13 subjects examined the anchor first and last, respectively. For the two-item bundles in the poor anchor condition, 22 and 10 subjects examined the anchor first and last, respectively. For the three-item bundles in the excellent anchor condition, 9 subjects examined the anchor first, 14 subjects examined the anchor second, and 8 subjects examined the anchor last. For three-item bundles in the poor anchor condition, 11 subjects examined the anchor first, 9 subjects examined the anchor second, and 12 subjects examined the anchor last.

[b] df=2,28

[c] df=2,29

[d] A linear contrast with weights 1, 0, and –1 was used to test the declining trend in the excellent anchor condition. The corresponding weights were –1, 0, and 1 in tests of the increasing trend in the poor anchor condition.

[e] df = 28.

[f] df = 29.

It is tempting to speculate why the bias due to insufficient adjustments was evident in the three-item bundles but not in the two-item bundles. First, recall that subjects determined the order of examination in study 1; in study 2, item descriptions were shown in a randomized order. A second point to note is subjects' tendency to engage in noncompensatory evaluation – three-item bundle evaluations were based largely on the two most important items.

Given the procedural difference between the two studies, it appears that implementing a noncompensatory evaluation may be easier in study 1 than in study 2. In study 1, subjects could implement a noncompensatory evaluation simply by terminating the evaluation after the first two most important items. How would a similar noncompensatory evaluation strategy operate in study 2? One way to process information noncompensatorily would be to consciously disregard evaluation of a less important item if encountered before the more important items. Or, the evaluation of all three items could be kept in short-term memory and then combined noncompensatorily. When compared with sequentially integrating information by means of anchoring and adjustment, both approaches appear either more unlikely or difficult -—especially in the case of three-item bundles. Therefore, subjects may be more likely to resort to a simplifying strategy such as anchoring and adjustment. However, the sequential nature of heuristic and the accompanying insufficient adjustments result in biased bundle evaluations.

4.2 Limitations and Research Opportunities

Caveats generally associated with laboratory experiments using student subjects are applicable here. Subjects were asked to evaluate bundles on the basis of information presented in several different paragraphs; no price information was made available. Naturally, the restricting nature of this information format limits the external validity of these results. More importantly, subjects were instructed to evaluate the bundle items one at a time. Therefore, the evaluation scenarios may have imposed a degree of structure to the evaluation process that is not typical of such evaluation situations. If information about all bundle items was presented together (say on one page of a catalog), it is possible strategies as well. For instance, subjects' attention may shift more frequently from one bundle item to another during the evaluation process. Hence, to some extent, evidence for the sequential nature of anchoring and adjustment may be overstated in this investigation.

It is also important to recognize the limitations of the methodological approach used in this investigation. As shown in Table 1, the research tradition within the area of anchoring and adjustment often uses outcomes to infer what the underlying process is – which is a possible concern, especially if outcomes inadequately capture the process. Although different methodological approaches were employed in the present research effort, the overall approach still reflects an emphasis on outcomes. For instance, the software monitored the information acquisition pattern of all subjects, but the protocol group was quite small ($n = 10$). The small size of the protocol group, coupled with the relative coarseness of the four coding categories, failed to yield a detailed process trace. Therefore, our understanding of bundle evaluation and of anchoring and adjustment could benefit from more process-oriented investigations. However, some researchers have argued that protocol analysis is likely to provide only limited additional insights about this specific process (see, e.g.; Davis et al. 1986). Given such concerns, developing innovative methodological approaches for exploring bundle evaluation and anchoring and adjustment should be a priority for future research in this area.

Finally, following Thaler (1985), one could conceptualize the overall value of a bundle as being composed of acquisition value and transaction value. Because price and nonprice information both play a role in the formation of acquisition and transaction values, the roles of both these types of information merit study. In this investigation, however, we focused only on the role of nonprice information. Empirical investigations are needed in which both price and nonprice information are manipulated in bundle offers to study their joint effects. More generally, additional research is needed to understand the role of nonprice and price information in the bundle evaluation process. Is the price information or the nonprice information examined first? If the prices of individual bundle items are provided along with the total bundle price, how is the evaluation process affected? Can a general anchoring and adjustment process be proposed to explain the processing of both price and nonprice information in bundle offers? Protocols, and computer software that can provide unobtrusive measures of the evaluation process, could be useful methodological tools for the first investigation of these issues.

References

Adams, W. J. and J. L. Yellen (1976). "Commodity Bundling and the Burden of Monopoly." Quarterly Journal of Economics, Vol. 90 (August), 475-498.

Anderson, N. H. (1981). "Foundations of Information Integration Theory." New York: Academic Press.

Biswas, A. and S. Burton (1993). "Consumer Perceptions of Tensile Price Claims: An Assessment of Claim Types Across Different Discount Levels." Journal of the Academy of Marketing Science, Vol. 21 (Summer), 217-230.

Block, R. A. and D. R. Harper (1991). "Overconfidence in Estimation: Testing the Anchoring and Adjustment Hypotheseis." Organizational Behavior and Human Decision Processes, Vol. 49 (August), 188-207.

Carlson B. W. (1990). "Anchoring and Adjustment in Judgements under Risk." Journal of Experimental Psychology, Vol. 16 (July), 665-676.

Dansby, R. E. and C. Conrad (1984). "Commoditiy Bundling." American Economic Review, Vol. 74 (May), 377-381.

Davis, H. L., S. J. Hoch and E.K. Easton Ragsdale (1986). "An Anchoring and Adjustment Model of Spousal Predictions." Journal of Consumer Research, Vol. 13 (June), 25-37.

Einhorn, H. J. and R. M. Hogarth (1975). "Unit Weighting Schemes for Decision Making." Organizational Behaviour and Human Performance, Vol. 13 (April), 171-192.

____ and R. M. Hogarth (1985). "Ambiguity and Uncertainty in Probabilistic Inference." Psychological Review, Vol. 92 (October), 461-465.

____, D. N. Kleinmuntz and B. Kleinmuntz (1979). "Linear Regression and Process-tracing Models of Judgment." Psychological Review, Vol. 86 (September), 465-485.

Ericsson, K. A. and H. A. Simon (1984). "Protocol Analysis." Cambridge, MA: MIT Press.

Friedlander, M. L. and S. J. Stockman (1983). "Anchoring and Publicity Effects in Clinical Judgment." Journal of Clinical Psychology, Vol. 39 (July), 637-643.

Gaeth, G. J., I. P. Levin, G. Chakraborty and A. M. Levin (1990). "Consumer Evaluation of Multi-Product Bundles: An Information Integration Analysis." Marketing Letters, Vol. 2 (1), 47-57.

Goldberg, S. M., P. E. Green and Y. Wind (1984). "Conjoint Analysis of Price Premiums for Hotel Amentities." Journal of Business, Vol. 57 (Suppl. 1, Part 2), S111-S132.

Guiltinan, J. P. (1987). "The Price Bundling of Services: A Normative Framework." Journal of Marketing, Vol. 51 (April), 74-85.

Hanson, W. and R. K. Martin (1990). "Optimal Bundle Pricing." Management Science, Vol. 36 (February), 155.174.

Hoffman, P. J. (1960). "The Paramorphic Represenattion of Clinical Judgment." Psychological Bulletin, Vol. 57 (2), 116-131.

Hogarth R. M. and H. J. Einhorn (1989). "Order Effects in Beleif Updating: The Belief Adjustment Model." Working paper, Center for Decision Research, University of Chicago.

Johnson, E. J. and D. A. Schkade (1988). "Bias in Utility Assessments: Further Evidence and Explanations." Management Science, Vol. 35 (April). 406-424.

Johnson, M. D. and C. P. Puto (1987). "A Review of Consumer Judgment and Choice." Review of Marketing, ed. Michael J. Houston, Chicago: American Marketing Association, 236-292.

Joyce, E. and G. C. Biddle (1981). "Anchoring and Adjustment in Probabilistic Inference in Auditing." Journal of Accounting Research, Vol. 19 (Spring), 120-145.

Kahneman, D. and A. Tversky (1979). "Prospect Theory: An Analysis of Decision Under Risk." Econometrica, Vol. 47 (March), 263-291.

Kirk, R. E. (1982). "Experimental Design: Procedures for the Behavioral Sciences." Pacific Grove, CA: Brooks/Cole.

Lopes, L. L. (1982). "Toward a Procedural Theory of Judgment." Working Paper 17, Wisconsin Human Processing Program." Departement of Psychology, University of Wisconsin, Madison.

____ (1985). "Averaging Rules and Adjustment Processes in Bayesian Inference." Bulletin of the Psychonomic Society, Vol. 23 (6), 509-512.

____ (1987). "Procedural Debiasing." Acta Psychologica, Vol. 64 (February), 167-185.

____ and Per-Hakan S. Ekberg (1980). "Test of an Ordering Hypothesis in Risky Decision Making." Acta Psychologica, Vol. 45 (August), 161-167.

Mano, H. (1990). "Anticipated Deadline Penalties: Effects on Goal Levels and Task Performance." In: Insights in Decision Making, ed. Robin M. Hogarth, Chicago: University of Chicago Press, 173-176.

Neter, J., W. Wasserman and M. J. Kutner (1990). "Applied Linear Statistical Model." 3rd ed., Homewood, IL: Irwin.

Northcraft, G. B. and M. A. Neale (1987). "Experts, Amateurs, and Real Estate: An Anchoring and Adjustment Perspective on Property Pricing Decisions." Organizational Behaviour and Human Decision Processes, Vol. 39 (February), 84-97.

Sawtooth Software (1987), Ci2 System, Version 2.0, Ketchum, ID: Sawtooth.

Schmalensee, R. (1984). "Gaussian Demand and Commodity Bundling." Journal of Business, Vol. 57 (Suppl. 1, Part 2). S211-S230.

Stigler, G. J. (1961). "United States v. Loews' Inc.: A Note on Block Booking." The Supreme Court Review, ed. Philip B. Kurland, Chicago: University of Chicago Press, 152-157.

Switzer, F. S. and J. A. Sniezek (1991). "Judgment Processes in Motivation: Anchoring and Adjustment Effects on Judgment and Behavior." Organizational Behaviour and Human Decision Processes, Vol. 49 (August), 208-229.

Thaler, R. (1985). "Mental Accounting and Consumer Choice." Marketing Science, Vol. 4 (Summer), 199-214.

Tversky, A. and D. Kahneman (1974). "Judgment under Uncertainty: Heuristics and Biases." Science, Vol. 185 (September), 1124-1131,

Venkatesh, R. and V. Mahajan (1993). "A Probabilistic Approach to Pricing a Bundle of Products and Services." Journal of Marketing Research, Vol. 30 (November), 494-508.

Wilson, L. O., A. M. Weiss and G. John (1990). "Unbundling of Industrial Systems." Journal of Marketing Research, Vol. 27 (May), 123-138.

Wright, W. F. and U. Anderson (1989). "Effects of Situation Familiarity and Financial Incentives on Use of the Anchoring and Adjustment Heuristic for Probability Assessment." Organizational Behavior and Human Decision Processes, Vol. 44 (August), 68-82.

Yadav, M. S. and K. B. Monroe (1993). "How Buyers Perceive Savings in a Bundle Price: An Examination of Bundle's Transaction Value." Journal of Marketing Research, Vol. 30 (August), 350-358.

Evaluating Multidimensional Prices in the Bundling Context

Andreas Herrmann[1], and Martin Wricke[2]

[1] **Andreas Herrmann**, Johannes Gutenberg-University of Mainz, Mainz, Germany.

[2] **Martin Wricke**, Johannes Gutenberg-University of Mainz, Mainz, Germany.

This paper is forthcoming in Pricing Strategy & Practice.

1 Introduction

Leasing offers, such as "... Jaguar XJ6 3.2, ..., down payment DM 22,325, DM 999 per month, 36 monthly installments ..." or "... Audi A4 Avant, ..., down payment DM 15,000, DM 157 per month, repayment period 36 months, residual payment DM 5,000 ...", have made regular appearances in advertisements for several years now. That offers of this kind are accepted by buyers of automobiles is confirmed by the leasing figures for German automobile manufacturers. Since the early 1990s, between 10 % and 28 % of all new automobiles have been sold under a leasing agreement, depending on the maker and the model series. The average leasing share for the automobile sector was approximately 23 % in Germany in 1995.

In view of the considerable marketing significance of leasing in many areas, manufacturers endeavor to structure their offers such that they appear as favorable as possible to potential buyers. Questions of the following kind arise frequently when these offers are elaborated: how does an individual integrate the different dimensions of an offer (down payment, monthly installments, repayment period in months) to form a global judgment? Which of these price dimensions are most important and which play a minor role in the consumers' decisions.

A brief glance at price theory literature provides us with an indication of the contribution made by the various schools of research towards answering these questions (Simon, 1989; Monroe, 1990). The theory of prices centers around normative approaches to pricing derived from the field of microeconomics, which attempt to maximize the economic target variables, such as turnover and profit. Taking the cost and price-demand functions and assumptions about the behavior of competitors as a basis, these models yield profit and turnover-maximization prices both for individual products and for the components of entire product lines. All these approaches share the methodological principle that consumers are economically rational.

However, the goal of behavioral science pricing models is to explain the actual, sometimes limitedly rational behavior of consumers when they attend to and process price information. The hypothetical constructs used to do so provide an indication of the activating and cognitive processes that take place in the consumers′ mind (Gurumurthy and Little, 1994). Among the constructs most relevant to the theory of prices are interest in the price, the price reasonableness rating and the value-for-money rating. Interest in the price is defined as the desire of a consumer to seek out price information and to take it into account in a purchase decision. Price judgment behavior embraces all the behavioral patterns which occur when price information is absorbed and processed. In contrast with interest in the price, it is the cognitive elements of the price behavior that are subsumed under this term, rather than the activating elements. A price reasonableness rating refers solely to the price level, in other words it takes no account of the quality of the offered commodity or of the scope of the services provided. A value-for-money rating, on the other hand, describes the price-performance ratio of the product.

All price theory approaches - whether derived from microeconomics or from behavioral science - are based at least implicitly on the assumption of a monodimensional pricing system. The manner in which multidimensional prices are judged, for example a price comprising down payment, monthly installment and repayment period in months dimensions, has not received attention from market researchers as a cognitive object. We shall thus attempt to answer the questions raised above by adopting the following approach: first of all, we shall explain information integration theory, which serves as the basis for all further observations. Four hypotheses describing the formation of a price judgment about multidimensional prices are then derived with the aid of this theory, and with reference to the relevant research literature. Finally, the formulated hypotheses are verified by means of an empirical study.

2 Essential Aspects of Information Integration Theory

Information integration theory is based on work by Anderson (1981, 1982) and Troutman and Shanteau (1976). The research they have conducted confirm the effectiveness of this approach for explaining the cognitive process of integrating information to form a global judgment. In marketing, information integration theory has been employed by Bettman, Capon and Lutz (1975) to analyze the development of attitudes and by Gaeth et al. (1991) to examine the intellectual processes involved in appraising product bundles.

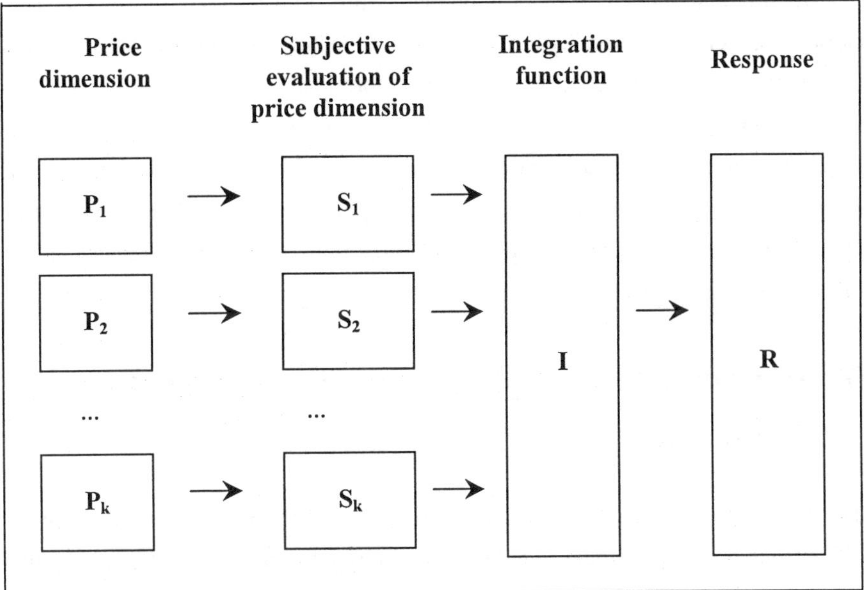

Fig. 1. The information integration process

The basic idea behind this approach can be explained with the aid of Figure 1 (Anderson, 1982). The starting point is stimuli P_1 to P_k, which in this case represent the levels of the different price dimensions (e.g., monthly installments DM 359, repayment period 36 months, down payment DM 5000). People transform the perceived levels of these stimuli into subjective partial ratings of dimensions that they have determined personally or that have been predefined by the researcher. If, as in our example, the judgment dimension is price reasonableness, the monthly installment is considered to be expensive and the down payment cheap, whereas the repayment period must first be imagined in terms of cheapness or expensiveness before a judgment can be formed. These partial ratings may then be combined to form a global judgment by means of an integration function and this global judgment is transformed into a concrete response (purchase or non purchase of the product in question).

Cognitive psychology analyses have revealed that, irrespective of the nature of the stimuli, human beings have a limited ability to process information. The storage capacity of short-term memory in particular is limited, so that any information that is not processed further is forgotten very quickly. In view of these limitations on information processing, it is also necessary to examine the steps a consumer follows to make a global judgment about a particular offer using the partial ratings of the perceived levels of the different price dimensions. Four hypotheses are developed to explain the formation of a price judgment about multidimensional prices.

3 Hypotheses for Explaining the Formation of a Price Judgment

The first hypothesis concerns the integration of the levels of the different price dimensions to form a global judgment. One normative model for integrating the price dimensions, of monthly installments, repayment period in months and down payment, corresponds to the formula used to determine the cash value of a series of payments:

$$\text{Cash value} = P + \sum_{t=1}^{N} \frac{I \cdot 12}{(1+i)^t} \qquad (1)$$

where:

P = down payment,

I = monetary amount of the monthly installments,

N = repayment period in years,

i = interest rate.

However, numerous mathematical operations would be necessary to form a judgment about an offer comprising, for example, monthly installments of DM 359, a repayment period of 36 months and a down payment of DM 5000 (Della Bitta, Monroe and McGinnis, 1981). Thus, is reasonable to suppose that the consumer simplifies the equation when appraising this offer, for example by not discounting the monthly installments. The following expression entails considerably less cognitive effort and serves as an approximate formula for the above equation:

$$\text{Cash value} = P + N \cdot I \cdot 12 \qquad (2)$$

Slovic (1972), moreover, provides evidence that predefined stimuli can generally be processed without any additional transformations (Jarvenpaa, 1989). According to this concreteness principle, a consumer integrates the levels of the different price dimensions (monthly installments, repayment period in months and down payment) to form a global judgment in a linear fashion, without taking into account either of the two equations (Bettman and Kakkar, 1977). This realization results in the first hypothesis:

Hypothesis 1: *The integration of the levels of different price dimensions can best be mapped with a linear model.*

The second hypothesis deals with the significance of the individual price dimensions for the global judgment. Studies by Tversky, Sattath and Slovic (1988) have revealed that a stimulus is easiest to process if it has the same unit (e.g., DM) as the target variable. The reason for this is readily apparent: a stimulus of this kind requires no mental transformations and entails considerably less cognitive effort. It is moreover possible to demonstrate that such stimuli are particularly important for the formation of a judgment. Individuals place more weight on them than on other stimuli, thereby elevating them to the status of central variables for the information processing process. Since we are concerned with consumers' rating of price reasonableness, there are grounds for suspecting that it is mainly the price dimensions expressed in DM on which this rating is based. This is expressed by hypothesis 2:

Hypothesis 2: *The down payment and monthly installments price dimensions have a greater influence on the formation of a price judgment than does the repayment period in months dimension.*

The third hypothesis concentrates on the question of the consistency of a price judgment. A price judgment is considered for our purposes to be consistent if the order of preference expressed by consumers for the various automobile offers corresponds to the order yielded by the cash value method. Fundamental work has been carried out on this subject by Hoffman and Blanchard (1961). According to their studies, judgment consistency declines as the amount of information to be processed increases (assuming that the quality of the information units remains the same). Johnson and Payne (1985) show that, when it comes to forming a price judgment, an individual finds it easier to determine the lowest offer if the price information is formulated simply. Hypothesis 3 thus states the following:

Hypothesis 3: *An increase in the number of price dimensions leads to a deterioration in the consistency of the price judgment.*

The fourth hypothesis attempts to explain the significance of key information for the formation of a price judgment. When evaluating a multidimensional phenomenon, an individual seeks to reduce the amount of cognitive effort to a minimum (Jacoby, Szybillo and Busato-Schach, 1977). Key information which integrates several other items of information is helpful in doing so. If the consumer is able to draw on information chunks, he or she will require less information to form a judgment. The effective annual interest rate could be used as key information in this example, as it integrates the levels of different price dimensions, such as monthly installments DM 359, repayment period 36 months, down payment DM 5000, in a single variable. Hypothesis 4 can thus be formulated as follows:

Hypothesis 4: *The consistency of the price judgment is improved if the effective annual interest rate is taken into account as key information.*

4 Empirical Studies

4.1 Study 1

The next step was to verify the above hypotheses concerning the formation of a price judgment about multidimensional prices with the aid of an empirical study. Various packages offered by an automobile manufacturer, each of which consisted of a fully equipped small automobile, were used for this purpose. The offered packages differed however with regard to the monthly installments, the length of the repayment period and the down payment. With two different monthly installments, two different repayment period lengths and two different down payment amounts, there is a total of eight different combinations of price dimension levels (see Table 1).

Table 1. Mean evaluation values for eight offers

Monthly Installment	Repayment Period	Down payment		Average	
		4000 DM	5000 DM	Installment	Period
629 DM	24	2.3 offer 1	4.2 offer 2	3.8	4.6 (24 Mon.)
	30	3.7 offer3	5.0 offer 4		
786 DM	24	4.9 offer 5	6.8 offer 6	6.4	5.6 (30 Mon.)
	30	6.3 offer 7	7.5 offer 8		
Average		4.3	5.9	1 = very cheap,..., 9 = very expensive	

In addition to specifying the independent variables (the three price dimensions), the objective was to operationalize the price reasonableness rating (the dependent variable). All previous findings relating to pricing policy confirm that this variable can best be measured on a scale from 1 (very cheap) to 9 (very expensive).

When the survey was conducted in May 1996, each of the 30 randomly selected respondents was presented with the eight offers and asked to appraise them individually with regard to the relevant target variable. An analysis of variance with repeated measures makes it possible to verify which rule has been chosen by the consumers for integrating the available information. Secondly, it also reveals the significance of the individual price dimensions for the global price judgment. Our initial interest is focused on the influence of the three price dimensions on the dependent variable. If this influence is non-existent, the target variable will have the same value in every cell. However, the mean values obtained for the cells (see Table 1) show that the mean scale value of the price reasonableness rating increases with the monthly installment amount, the length of the repayment period and the down payment. Whereas for offer 1, for example (monthly installments DM 629, repayment period 24 months, down payment DM 4000), the mean value is 2.3, the corresponding value for offer 8 (monthly installments DM 786, repayment period 30 months, down payment DM 5000) is 7.5.

Table 2. Results of the F-test for three dimensional offers

Price dimension	Degrees of freedom	F-value	Pr > F
Monthly installment (I)	1	72.43	.00
Down payment (P)	1	49.67	.00
Repayment period (R)	1	38.43	.00
I × P	1	1.72	.21
I × R	1	0.59	.53
P × R	1	0.12	.74
I × P × R	1	1.03	.46
Residual	225		

The omnibus F-test can be used to determine the statistical significance of the main and interaction effects. Table 2 shows that the empirical F values of the three main effects are statistically significant. The monthly installments, the length of the repayment period and the down payment each have a statistically significant influence on the respondents′ price reasonableness ratings. Moreover, all interaction terms between the three price dimensions are not statistically significant.

The results of the hypothesis test reveal that the interaction terms are not necessary to map the information integration process. Thus, the linear model appears to be a suitable means for representing the formation of a price reasonableness rating concerning an offer comprising monthly installments, a repayment period in months and a down payment. A glance at Figure 2, which shows the mean values previously obtained in the cells of the data matrix, is sufficient to confirm this. The roughly parallel gradients of the curves on the various graphs clearly illustrate the relevance of the linear model for representing the process of information integration. With monthly installments of DM 629, for example, a change in the price reasonableness rating is more or less independent of the down payment, owing to the simultaneous change in the length of the repayment period (see Fig. 2). This result would initially appear to confirm hypothesis 1.

It is necessary to consider questions of the following kind in order to verify hypothesis 2: does the price reasonableness rating of individuals who are presented with an offer with monthly installments of DM 629 differ from that of consumers offered monthly installments of DM 786? Or: are the respondents whose offer includes a repayment period of 24 months characterized by a different mean scale value for price reasonableness than the interviewees confronted with a repayment period of 30 months? Since the difference between the two mean values of a price dimension, in other words between 4.3 and 5.9 or 3.8 and 6.4 or 4.6 and 5.6 (see Table 1), deviates significantly from zero, the following conclusions can be drawn.

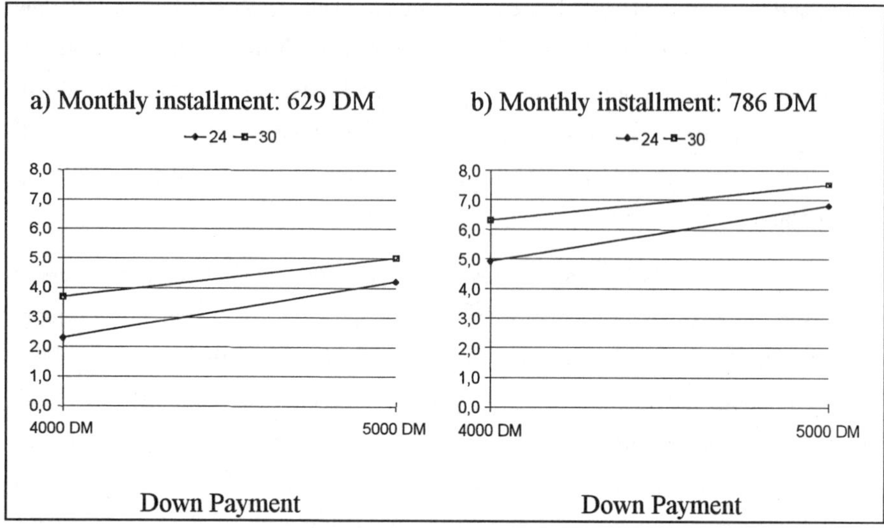

Fig. 2. Relationship between monthly installment, down payment, and repayment period

The monthly installment effect size, ω^2, explained the greatest percentage of variance (23 percent), followed by the down paymant effect size (18 percent) and the repayment period effect size (15 percent). So hypothesis 2 is also confirmed.

4.2 Study 2

Various packages offered by an automobile manufacturer, each of which consisted of a fully equipped small automobile, were used once again to test hypothesis 3. In the first case (study 2a), the offers differ merely with regard to the monthly installments and the length of the repayment period. With three different monthly installments and three different repayment periods, there are nine possible combinations of price dimension levels. In the second case (study 2b), the offers deviate from one another with respect to the monthly installments, the repayment period in months, the down payment and the residual payment. With two levels per price dimension, a total of 16 different offers are available. Like the empirical study described above, each of these two studies entailed prompting 30 respondents to give their price reasonableness ratings for the various automobile packages on offer on a scale from 1 to 9. The interviewees were also asked to specify what they thought to be the market interest rate on a loan of DM 25,000.

A hypothesis test method initially revealed the statistical significance of the two or four main effects (in other words the price dimensions); the first to third-order interaction terms do not differ significantly from zero (see Tables 3 and 4). The assumption that the price dimensions are integrated in a linear fashion to form the price reasonableness rating apparently also applies to these two studies. The effect size of monthly installment and repayment period in study 2a was 26 percent and 19 percent, respectively. In study 2b the ω^2 is 19 percent for monthly installment, 16 persent for down payment, 13 percent for residual payment and 10 percent for repayment period. All the offers were then discounted at the market interest rates specified by the respondents. The calculated cash values can be arranged in a rank order and compared with the order derived from the ratings. The coefficient of correlation is 0.78 for two price dimensions, 0.70 for three dimensions and 0.68 for four dimensions. Since the consistency of the judgments declines as the number of price dimensions increases, hypothesis 3 can also be considered to be confirmed.

Table 3 Results of the F-Test for the two dimensional offers

Price dimension	Degrees of freedom	F-value	Pr > F
monthly installmant (I)	1	38.86	.00
repayment period (R)	1	26.13	.00
I × R	1	0.47	.83
Residual	113		

Table 4 Results of the F-Test for the four dimensional offers

Price dimension	Degrees of freedom	F-value	Pr > F
monthly installment (I)	1	108.65	.00
down payment (P)	1	87.23	.00
repayment period (R)	1	65.29	.00
residual payment (A)	1	52.14	.00
I × P	1	1.46	.28
I × R	1	.28	.65
I × A	1	1.74	.24
P × R	1	.67	.57
P × A	1	1.43	.39
R × A	1	1.13	.42
I × P × R	1	.86	.53
I × P × A	1	.23	.72
P × R × A	1	.35	.65
I × P × R × A	1	.11	.86
Residual	450		

4.3 Study 3

Hypothesis 4 was verified by adding offer-specific information about the effective annual interest rate (see Table 5) to the experimental layout made up of the three price dimensions (see Table 1). A fictitious automobile value of DM 19,000 was taken as the cash value. Once again, 30 respondents were asked to rate the offers presented to them with regard to their price reasonableness. It was striking that only 11 out of the total of 30 interviewees were familiar with the concept of an effective interest rate or at the very least took it into consideration as a decision criterion. As mentioned earlier, the automobile offers can be arranged in a rank order on the basis of the ratings and compared with the rank order derived from the effective interest rate. The coefficient of correlation for those respondents who considered the effective interest rate to be key information is 0.93, while the corresponding value for the other interviewees is 0.65. Since the consistency of the judgments increases if the effective annual interest rate is taken into account as key information, hypothesis 4 is likewise confirmed.

Table 5 Effective annual interest rate as key information

Monthly installment	Repayment period	Down payment	
		4000 DM	5000 DM
629 DM	24	offer 1 (1%)	offer 2 (7%)
	30	offer 3 (18%)	offer 4 (24%)
786 DM	24	offer 5 (23%)	offer 6 (30%)
	30	offer 7 (38%)	offer 8 (45%)

5 Summary

The central theme of this article is the derivation of four hypotheses for explaining the formation of a price judgment about multidimensional prices, together with an empirical study designed to verify these hypotheses. The following results were obtained:

1. The integration of individual price dimensions to form a global price judgment can best be mapped with a linear model.
2. The down payment dimension and the monthly installments dimension have a greater effect on the formation of a price judgment than the repayment period in months dimension.
3. An increase in the number of price dimensions leads to a deterioration in theconsistency of the price judgment.
4. The consistency of the price judgment is improved if the effective annual interest rate is taken into account as key information.

These results provide managers with numerous clues regarding the ideal price structure of a product. The aim of the study, however, was not so much to determine the optimum format for presenting price information about a specific product from a behavioral science point of view, but rather to demonstrate how the process by which a price reasonableness judgment is formed can be analyzed when a multidimensional pricing system is used. In a concrete situation, greater importance would need to be attached to making the offers realistic as well as to legal requirements. In addition, further analyses would be necessary to establish the relationship between the offers considered by consumers to be the most favorable and the one they actually choose.

References

Anderson, N. H. (1981). Foundations of Information Integration Theory, McGraw-Hill, New York, NY.

Anderson, N. H. (1982). Methods of Information Integration Theory, McGraw-Hill, New York, NY.

Bettman, J., N. Capon and H. Lutz (1975). "Cognitive Algebra in Multi-Attribute Attitude Models." Journal of Marketing Research, Vol. 12 (May), 151-64.

Bettman, J. and P. Kakkar (1977). "Effects of Information Presentation Format on Consumer Information Acquisition Strategies." Journal of Consumer Research, Vol. 3 (March), 233-40.

Della Bitta, A., K. B. Monroe and J. McGinnis (1981). "Consumer Perceptions of Competitive Price Advertisements." Journal of Marketing Research, Vol. 18 (November), 416-27.

Gaeth, G. J., I. P. Levin, G. Chakraborty and A. M. Levin (1990). "Consumer Evaluation of Multi-Product Bundles: An Information Integration Analysis." Marketing Letters, Vol. 2 (January), 47-57.

Gurumurthy, K. and J. Little (1994). "An Empirical Analysis of Latitude of Price Acceptance in Consumer Package Goods." Journal of Consumer Research, Vol. 21 (December), 408-18.

Hoffman, P. J. and W. A. Blanchard (1961). "A Study of the Effects of Varying Amounts of Predictor Information on Judgement." Working paper, University of Oregon, Portland, Or.

Jacoby, J., G. J. Szybillo and J. Busato-Schach (1977). "Information Acquisition Behavior in Brand Choice Situations." Journal of Consumer Research, Vol. 3 (March), 209-16.

Jarvenpaa, S. L. (1989). "The Effect of Task Demand, and Graphic Formation Information Processing Strategies." Management Science, Vol. 35 (March), 285-303.

Johnson, E. and J. Payne (1985). "Effort and Accuracy in Choice." Management Science, Vol. 31 (April), 395-414.

Monroe, K.B. (1990). Pricing: Making Profitable Decisions. 2nd ed., McGraw-Hill, New York, NY.

Simon, H. (1989). Price Management. North-Holland, Amsterdam.

Slovic, P. (1972). "From Shakespeare to Simon: Speculations and some Evidence about Man`s Ability to Process Information." Oregon Research Institute Monograph, Vol. 12, 10-23.

Troutman, M. C. and J. Shanteau (1976). "Do Consumers Evaluate Products by Adding or Averaging Attribute Information." Journal of Consumer Research, Vol. 3 (September), 101-6.

Tversky, A., S. Sattath and P. Slovic (1988). "Cognitive Weighing in Judgement and Choice." Psychological Review, Vol. 95, 371-84.

Product and Service Bundling Decisions and their Effects on Purchase Intention

Andreas Herrmann[1], Frank Huber[2], and Robin Higie Coulter[3]

[1] **Andreas Herrmann**, Johannes Gutenberg-University of Mainz, Mainz, Germany.

[2] **Frank Huber**, Business School at the University of Mannheim, Mannheim, Germany.

[3] **Robin Higie Coulter**, Department of Marketing, University of Connecticut, Storrs, Connecticut

This paper was published in Pricing Strategy & Practice, 1997, pp. 99-107.

1 Introduction

Bundling of products, product components, and services is an important consideration for manufacturers, retailers, and service providers bringing their goods and services to market. Bundling typically takes one of two forms: pure or mixed (Adams and Yellen, 1976). Pure bundling refers to a strategy in which only a bundle of items or components is available for purchase; in other words, buyers must purchase the bundle, they do not have the option of purchasing individual components. In contrast, mixed bundling gives buyers the option of purchasing either the bundle, or any or all of the individual components.

To differentiate their products and services from the competition, manufacturers, retailers, and service providers often use bundling strategies (e.g., automobile option packages, stereo/compact disc/tape deck packages, travel packages that vary in their comprehensiveness of room, board and entertainment coverage, and automobile service centers that offer packages that vary on their maintenance coverage). Recent work on bundling has provided important insights with regard to bundling (Guiltinan, 1987), consumers' evaluation of multi-product bundles (Gaeth et al., 1990), their perceptions of savings when they evaluate bundle offers (Yadav and Monroe, 1993), and their evaluation of bundles that include different anchor products, as well as different numbers of products (Yadav, 1994). Nonetheless, many questions about consumers' evaluations of product and service bundles remain unanswered. Of note, Yadav and Monroe (1993) suggest future research should focus on the joint effects of price and non-price information in consumers' bundle evaluations.

The purpose of this paper is to examine four factors expected to affect consumers' intentions to purchase product and service bundles. These factors include: whether the bundle is pure or mixed, the price discounts of a pure bundle in comparison to the sum of the components of a mixed bundle, the functional complementarity of the components in the bundle, and the number of components in the bundle. Specifically, we are interested not only in the effects of each factor on purchase intention, but also the combined effects of these factors. We begin by reviewing these factors as related to bundling and intention to purchase, and then, describe a study of bundling in the context of automobile choice and automotive service choice. Our results provide implications for managerial decisions related to bundle pricing and composition.

2 Pricing Pure and Mixed Bundles

The typical pricing strategy with bundles is to offer a pure bundle at a discount as an incentive for consumers to purchase a package rather than to purchase individual components of the package. Consider the following example. An automotive manufacturer offers a sporty package (e.g., a three pronged steering wheel, sporty seats, tachometer, and four aluminum wheel rims) as a pure bundle for \$2,400, or as a mixed bundle, in which case the sum of the components is \$3,000. In the former case, the consumer receives all of the components at a 20 percent discount; in the latter case, the individual can choose from the components in the bundle (e.g., select only the tachometer and the aluminum wheel rims) but will pay a higher price per component than would be the case if he purchased the pure bundle.

Research suggests that consumers tend to use the individual component prices for a mixed bundle as their reference price in judging the value of a pure bundle that includes the same items (Yadav and Monroe, 1993). Thus, consumers perceive the pure bundle as providing more value for the dollar than the mixed bundle, and hence are more likely to purchase the pure than the mixed bundle.

3 Number of Components in a Bundle

The number of components is another concern to those responsible for constructing bundles (Ansari, Siddarth and Weinberg, 1996). Lawless (1991) argued that, from a strategic perspective, the more products (or services or components) that a firm includes in a bundle, the more difficult it is for the competition to duplicate the bundle. Nonetheless, from an information processing perspective, it is important to consider just how many bundle components consumers can process and/or factor into their decisions. Research suggests that consumers process and value information about a set of attributes, until the amount of information exceeds their cognitive capacity. This line of reasoning suggests that more components in a bundle is better, i.e., results in greater purchase intention, until the number exceeds processing capacity, at which point information overload occurs and purchase intention decreases.

4 Functional Relationship Among Bundle Components

Identification and inclusion of the "optimal" set of components in a bundle is another key concern of manufacturers, retailers and service providers. Should a multi-component bundle consist of functionally related, complementary attributes (e.g., a centralized lock system and an alarm system) or functionally unrelated attributes (e.g., a centralized lock system and a sunroof)? From the seller's perspective, complementary bundle components simplify cross-selling, post-sales support, and potentially increase consumer loyalty (Lawless, 1991; Paun, 1993). Gaeth et al. (1990) found that consumers evaluate bundles consisting of functionally related products differently than they evaluate bundles consisting on functionally unrelated products.

In our context, one might argue that upon scanning components of a bundle, consumers will perceive a bundle that has complementary attributes as more favorable (and hence, more willing to purchase) than a bundle that is comprised of functionally unrelated attributes.

5 Methodology

We conducted two studies, one in the context of a product choice (i.e., likelihood of buying a car) and the other in the context of a service choice (i.e., likelihood of purchasing an automotive maintenance service package). The components under consideration were the same for the product and service experiments, they included: type of bundling (2 levels: pure or mixed), price discount for the pure bundle in comparison to the sum of the prices for the individual components (3 levels: 0%, 10%, 20%), functional complementarity of the components (3 levels: very complementary, somewhat complementary, not at all complementary), and number of components or elements included in the bundle (3 levels: 3, 5, 7).

We conducted two (2 x 3 x 3 x 3) full factorial experiments in November and December 1994 in Munich, Germany. A total of 540 subjects (car owners who were in the market to buy a new car) from a panel participated in the study. The sample, in terms of socioeconomic variables, was representative of the German population. Each person was randomly assigned to one of the fifty-four cells (ten subjects per cell) in each experiment; thus, it was unlikely that the subject was assigned to the same cell in both experiments. Experiment 1 assessed purchase likelihood for an automobile and Experiment 2 assessed purchase likelihood for an automotive service package. Experiment 1 preceded Experiment 2 for all subjects.

Based on the cell to which the subjects were assigned in Experiment 1, they received a sheet that included information on each of the factors we manipulated: type of bundle, price discount, functional complementarity and number of components. Each description contained information about a standard automobile completely assembled by a German car manufacturer and a picture of the car's exterior; this part of the description was constant across all 54 cells.

Information about Product Bundles			
Standard Model 1[a]	**Standard Model 2**	...	**Standard Model 54**
Standard equipped German manufacturer's car model **Bundle** Automatic Locking System Alarm system Passenger Airbag Price: DM 2,130 **Price for Individual Items** Auto Locking System DM 750 Alarm system DM 530 Passenger Airbag DM 850 Total for Individual Items: DM 2,130 **Components are available only in the bundle**	Standard equipped German manufacturer's car model **Bundle** Automatic Locking System Alarm system Passenger Airbag Price: DM 2,130 **Price for Individual Items** Auto Locking System DM 750 Alarm system DM 530 Passenger Airbag DM 850 Total for Individual Items: DM 2,130 **Components are not available in the bundle, only individually**	...	Standard equipped German manufacturer's car model **Bundle** Radio Metallic Paint Passenger Airbag Sun roof Automatic Locking System Alarm system Velour seats Price: DM 5,264 **Price for Individual Items** Radio DM 740 Metallic Paint DM 1,260 Passenger Airbag DM 850 Sun roof DM 1,490 Auto Locking System DM 750 Alarm system DM 530 Velour seats DM 960 Total for Individual Items: DM 6,580 **Components are not available in the bundle, only individually**

[a] Standard Model 1 represents a pure bundle at no discount with three "very related" (security/safety) components.

Fig. 1. Examples of product bundle composition

(The exact description of the car is not included in this document, due to the proprietary nature of the research.) The description contained two packages, one a pure bundle (with 3, 5, or 7 components) with a quoted price (i.e., the 0%, 10%, or 20% discount compared to the sum of the individual components), and one a mixed bundle (with the same number of attributes as the pure bundle) with the respective prices of the component parts. Functional complementarity was operationalized by designing bundles that had functionally related attributes (e.g., the safety package included: a centralized security system, an alarm system and a passenger-side airbag), and bundles that included unrelated attributes (e.g., a centralized security system, a sun roof and aluminum wheel rims). Additionally, a statement about whether the components could be purchased only as a pure bundle or were available individually (i.e., as a mixed bundle) was included on the information sheet.

Figures 1 and 2 illustrate three of the fifty-four stimuli for Experiment 1 (the automobile) and Experiment 2 (the service package), respectively. After reviewing the information sheet, subjects indicated their intention to purchase the described bundle on a seven point scale in which 1 is "Not at all likely to purchase" and 7 is "Very likely to purchase."

Information about Service Bundles			
Service Bundle 1	**Service Bundle 2**	...	**Service Bundle 54**
Oil change Brake and brake fluids test Battery test Price: DM 320	Oil change Brake and brake fluids test Battery test Price: DM 320		Oil change Brake and brake fluids test Battery test CO2 test Fan belt test Change tires Rotate tires Price: DM 416
Price for Individual Items Oil change DM 120 Brake/brake fluids test DM 140 Battery test DM 60 Total for Individual Services: DM 320 **Services are available only in the bundle**	**Price for Individual Items** Oil change DM 120 Brake/brake fluids test DM 140 Battery test DM 60 Total for Individual Services: DM 320 **Services are not available in the bundle, only individually**	...	**Price for Individual Items** Oil change DM 120 Brake/brake fluids test DM 160 Battery test DM 30 CO2 test DM 90 Fan belt test DM 20 Change tires DM 70 Rotate tires DM 30 Total for Individual Services: DM 520 **Services are not available in the bundle, only individually**

[a] Service Bundle 1 represents a pure bundle at no discount with three "very related" ("regular maintenance") services.

Fig. 2. Examples of service bundle composition

6 Findings

6.1 Individual Effects

Our results indicate main effects for each of the four bundle factors for both product and service purchase intention. First, we found a price discount main effect for both the automobile ($F_{2/486}$=138.87; p<.001) and the automotive service ($F_{2/486}$=35.37; p<.001). As might be expected, as the price discount for the bundle increased from 0% to 20%, consumers were more likely to report an intention to purchase the product or service bundle (Table 1 provides the purchase intention means for the independent variables). In the product context, the price discount effect size, ω^2, explained the greatest percentage of variance (28%); in the service context, ω^2 was 9%.

Table 1. Mean intention to purchase values for main effect results[a]

Manipulated Variable	Automobile Bundle	Automotive Service Bundle
Price Discount		
0 percent	2.69 [a,b]	2.67 [a,b]
10 percent	3.24 [a,c]	3.14 [a,c]
20 percent	4.29 [b,c]	3.62 [b,c]
Bundle		
Pure	3.59	3.30
Mixed	3.23	2.99
Number of Components		
Three	3.36 [a]	2.99 [a]
Five	3.71 [a,b]	3.51 [a,b]
Seven	3.16 [b]	2.93 [b]
Functional Complementarity		
Related Components	3.91 [a,b]	3.69 [a,b]
Somewhat Related	3.29 [a]	3.12 [a,c]
Not At All Related	3.02 [b]	2.62 [b,c]

[a]*Note*: Mean values are based on a 1 to 7 scale in which 1 is "Not at all likely to purchase" and 7 is "Very likely to purchase." Within columns (i.e., for the product or for the service) and by manipulated variables, means with the same superscript differ significantly based on Scheffé comparisons (p<.05). For example, with regard to the automobile bundle and the price discount manipulated variable, the 0% discount resulted in significantly lower purchase intention than the 10% price discount (represented by the [a]) and the 20% price discount (represented by the [b]), and the 10% discount resulted in significantly lower purchase intention than the 20% price discount (represented by the [c]).

As research suggests, we found that a pure bundle resulted in greater purchase intention than a mixed bundle for both the automobile ($F_{1/486}$=20.34; p<.001) and automotive service ($F_{1/486}$=11.25; p<.01). The ω^2 in the product and service contexts was 2% and 1%, respectively.

Our results indicated a main effect for number of components in the bundle for the automobile ($F_{2/486}$=16.29; p<.001) and the automotive service ($F_{2/486}$=15.80; p<.001). The ω^2 for the product and service bundles, respectively, was 3% and 4%. For the automobile, we found a curvilinear relationship; the quadratic term was significant ($F_{1/537}$=15.99; p<.001), as was the contrast between five attributes and three and seven attributes (t_{537}=4.00; p<.001). The automotive service results were similar; the quadratic term ($F_{1/537}$=23.04; p<.001) and the contrast (t_{537}=4.80; p<.001) were significant. As shown in Figure 3, purchase intention was greatest for both the automobile and the automotive service when the bundle had five, rather than three or seven, components.

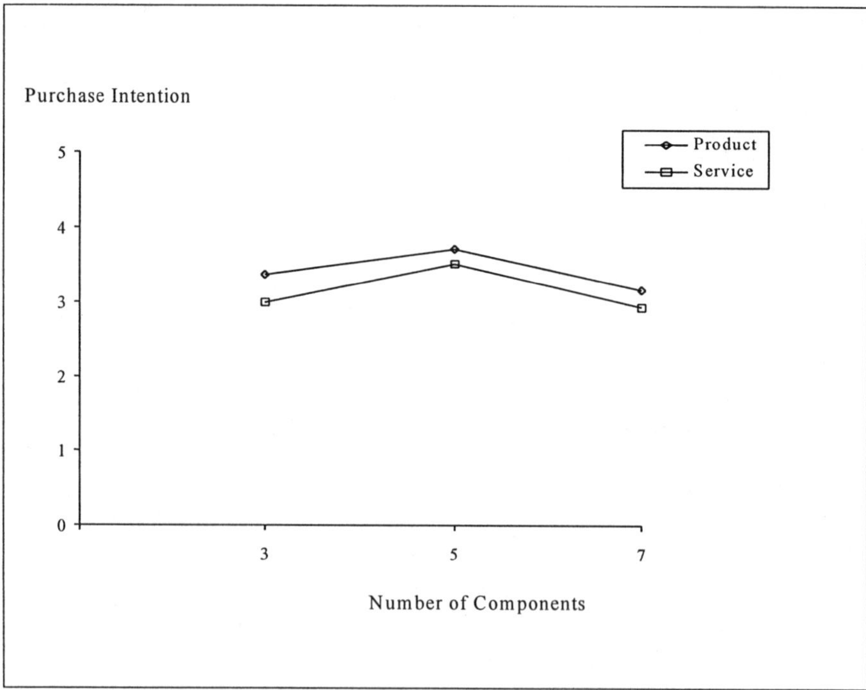

Fig. 3. Purchase intention means for the number of components in product and service bundles

Also, as might be expected, the greater the functional complementarity among the components or elements of a bundle, the greater the likelihood of purchase of the automobile ($F_{2/539}$=44.17, p<.001) and the automotive service ($F_{2/539}$=44.62; p<.001). As the relationship among the components increased from "not at all related," to "somewhat related," to "very related," intention to purchase also increased. Again, Table I shows the purchase intention means for the three levels of functional complementarity; "very related" components generated significantly greater purchase intention than the other two manipulations. Functional complementarity, with an effect size of 11%, explained the greatest percentage of variance in the automotive service purchase intention, and accounted for 9% of the variance in the automobile purchase intention.

6.2 Interaction Effects

In addition to the main effects, we found three significant interaction effects for the product and service choices. First, we found a price discount by functional complementarity interaction for both the automobile ($F_{4/539}$=7.23; p<.001) and the automotive service ($F_{4/539}$=4.86; p<.01) purchase intention. In general, the findings indicated that the more functionally compatible the components and the greater the price discount, the greater the intention to purchase. Post hoc Scheffé analyses (p<.05) for the automobile indicated and as Figure 4 illustrates, mean purchase intention for "very related" bundle components priced at a 20% discount ($\bar{x} = 5.1$) was significantly greater than purchase intention for other complementarity-price discount combinations. For the automotive service, the "very related" component bundle priced at a 20% discount generated a significantly greater purchase intention ($\bar{x} = 4.3$) than the "not related" and "somewhat related" functional components priced at no discount and at a 10% discount. In other words, for "very related" service bundle components, price discounts of 10% and 20% did not generate significantly greater purchase intention than was generated by no price discount (see Figure 4).

Second, we found a functional complementarity by bundle interaction for the automobile ($F_{2/539}$=6.66; p<.001) and the automotive service ($F_{2/539}$=3.11; p<.05) purchase intention. When the components were most functionally complementary, consumers were more likely to purchase the pure than the mixed bundle. For both product and service, post hoc Scheffé analyses ($p<.05$) indicated that the pure bundle with "very related" components generated greater purchase intention than each of the other bundle-functional complementarity combinations.

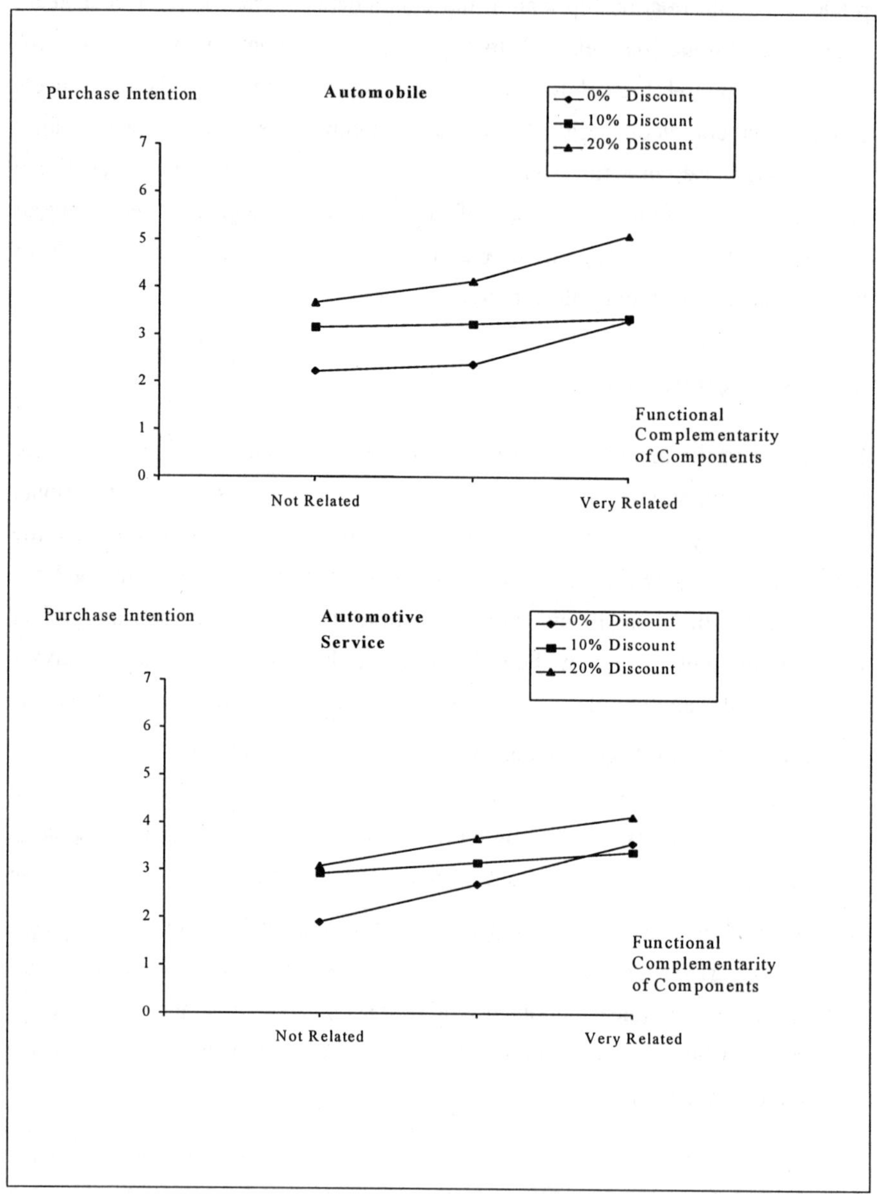

Fig. 4. Purchase intention means for the functional complementarity by price discount interaction

Third, we found a price discount by number of components interaction ($F_{4/539}$=4.08; p<.01) for the automobile purchase intention. Post hoc Scheffé analyses (p<.05) suggested that bundles with three, five or seven components at a 20% discount generate the same level of purchase intention, and that they generate greater purchase intention than bundles (regardless of the number of components) priced at a 10% or no discount.

Finally, our results for the automotive service bundle indicated an interaction between the number of components and functional complementarity ($F_{4/539}$=4.50; p<.001). Post hoc Scheffé analyses (p<.05) indicated that five very complementary components resulted in greater purchase intention than any other number of components-functional complementarity combination.

7 Managerial Implications and Recommendations

Our findings suggest that manufacturers, retailers, and service providers considering a bundling strategy should pay attention to the four variables we considered, since each had an effect on purchase intention, and some interactions were present. Our findings were relatively consistent across product (automobile choice) and service (automotive service choice) contexts, and in general, they illustrated that pure bundles are preferred to mixed bundles, and a greater price discount is preferred to a lesser one. Our results also indicated that five component bundles generate greater purchase intention than either three or seven component bundles, and that "very related" bundle components result in greater purchase intention than either moderately or not related components. For the automobile, price discount explained 28 percent of the variance in purchase intention, and functional complementarity explained nine percent. For the automotive service, the functional complementarity of components explained 11% of the variance and price discount explained nine percent.

It is important to note that these individual effects are tempered in the context of some interaction effects which are discussed in the next two sections. Additionally, although our results do not indicate a four-way interaction among the variables under investigation, a pure bundle with a 20% discount that includes five "very related" components achieved the highest purchase intention (on a seven-point scale) for the automobile ($\bar{x}$ = 6.3) and the automotive service ($\bar{x}$ = 5.9). Other product combinations with the purchase intention greater than or equal to 5 included: 1) a pure bundle with seven "very related" components at a 20% discount, 2) a pure bundle with three "very related" components at a 20% discount, and 3) a mixed bundle with five "very related" components at a 20% discount. No other service combination had a purchase intention with a scale value of five or more.

7.1 Product Bundling

Our results suggest that manufacturers and retailers should be sensitive to interaction effects between bundle factors, particularly with regard to the price discount. First, managers need to consider the functional complementarity of bundle components when establishing the price discount. For all levels of functional complementarity (very, somewhat and note related), a 20% discount resulted in greater purchase intention than no discount or a 10% discount. For "very related" component bundles, there was no difference in the purchase intention generated by no discount and a 10% discount. For moderately and not related component bundles, however, the 10% discount generated greater purchase intention than no discount. Thus, in the former case involving "very related" components, offering a 10% discount (rather than no discount) does not result in increased purchase intention, whereas for somewhat and not related components, a 10% discount (compared to no discount) can affect purchase intention.

The price discount by number of components interaction also has pricing implications. Regardless of the number of component (three, five or seven) in a bundle, a 20% discount resulted in a greater purchase intention than no discount or a 10% discount.

Finally, when constructing bundles designed to generate purchase intention, managers should be cognizant that a pure bundle with "very related" components resulted in a greater purchase intention than other functional complementarity-bundling combinations.

7.2 Service Bundling

Our results also indicate that service providers should be attentive to interaction effects between bundle factors, particularly those with regard to functional complementarity. First and similar to the product context finding, service providers need to consider the functional complementarity of bundle components when establishing the price discount. For "very related" component bundles, although a 20% discount results in the greatest purchase intention, there is no significant difference in the purchase intention generated by no discount, a 10% discount and a 20% discount. For less compatible components, there is no difference in the purchase intention generated by a 20% and 10% discount in the service context, and both generate greater purchase intention than no discount. Based on these findings, service providers must carefully assess the value of pricing "very related" service bundles at a discount, and are best advised to price "somewhat related" and "not at all related" service bundles at a 10% discount.

Two other functional complementarity interaction effects (with the bundle and the number of components) have implications for service providers' bundle construction. Similar to the product findings and related implications, service providers should establish pure service bundles with "very-related" components (rather than other functional complementarity-bundling combinations) to achieve greater purchase intention. Further, constructing a bundle with five "very related" components should achieve greater purchase intention than other functional complementarity-number of components combinations.

8 Summary

This study examined the individual and combined effects of four bundle factors on product and service purchase intention. Price discount and complementarity of bundle components appear to be key drivers of purchase intention for both the automobile and automotive service bundles. Moreover, the interaction between the two factors suggests that there is a threshold level of discounting necessary to affect purchase intention. On average, a 20% discount results in the greatest purchase intention, but care must be given to assessing the value of a 10% discount. In some cases (somewhat and not related component service bundles), the 10% discount will save money because it generates the same level of purchase intention as a 20% discount. In other instances (very related component product and service bundles), a 10% discount results in the same purchase intention as no discount, and therefore would be an inefficient use of resources. In the last case (somewhat and not related component product bundles), a 10% discount results in greater purchase intention than no discount but less purchase intention than a 20% discount. In general, our findings emphasize the need to attend to not only to individual bundle variable effects but also the interaction effects.

Future research should continue to investigate the individual and combined effects of bundle components for other service and product choices. Studies might also examine bundle factor effects not only on purchase intention, but also on consumer information processing variables of interest. Additionally, the effects of these bundle factors might be investigate with regard to consumer product and service loyalty, as well as with regard to strategies for effectively communicating bundle information to consumers.

References

Adams, W. J. and J. L. Yellen (1976). "Commodity bundling and the burden of monopoly." Quarterly Journal of Economics, Vol. 90 (August), 475-89.

Ansari, A., S. Siddarth and C. B. Weinberg (1996). "Pricing a bundle of products or services: the case of nonprofits." Journal of Marketing Research, Vol. 33 (February), 86-93.

Gaeth, G. J., I. P. Levin, G. Chakraborty and A. M. Levin (1990). "Consumer evaluation of multi-product bundles: an information integration analysis." Marketing Letters, Vol. 2 (January), 47-57.

Guiltinan, J. P. (1987). "The price bundling of services: A normative framework." Journal of Marketing, Vol. 51 (April), 74-85.

Lawless, M. W. (1991). "Commodity bundling for competitive advantage: strategic implications." Journal of Management Studies, Vol. 28 (May), 267-80.

Paun, D. (1993). "When to bundle or unbundle products." Industrial Marketing Management, Vol. 22 (February), 29-34.

Yadav, M. S. (1994). "How buyers evaluate product bundles: a model of anchoring and adjustment." Journal of Consumer Research, Vol. 21 (September), 342-53.

Yadav, M. S. and K. B. Monroe (1993). "How buyers perceive savings in a bundle price: an examination of a bundle's transaction value." Journal of Marketing Research, Vol. 30, (August), 350-8.

Utility-Oriented Design of Service Bundles in the Hotel Industry, Based on the Conjoint Measurement Method

Hans H. Bauer[1], Frank Huber[2], and Richard Adam[3]

[1] **Hans H. Bauer**, Business School at the University of Mannheim, Mannheim, Germany.
[2] **Frank Huber**, Business School at the University of Mannheim, Mannheim, Germany.
[3] **Richard Adam**, Arabella Hotel, Munich, Germany.

1 On the Significance of Product Bundles for a Company

When it comes to designing a marketable service, an increasing number of suppliers meanwhile combine their products in a package, tailored to a specific application, and then offer this package at a bundle price (Adams and Yellen, 1976; Simon, 1992; Diller, 1993; Nieschlag, Dichtl and Hoerschgen, 1994). Price bundling is particularly relevant in connection with the operational services offered within the hotel industry (Goldberg, Green and Wind, 1984; Guiltinan, 1987; Bojanic, 1988). By browsing through the advertisement columns of pertinent newspapers, magazines or travel brochures, a person seeking a break away from home will frequently encounter offers of overnight accommodation where the hotel management charges a flat-rate price to the consumer for a stay in a specific category of room, including the use of various fitness amenities available in the hotel as well as catering services.

This paper attempts to demonstrate how a hotel manager can make use of the conjoint analysis method to solve the typical problems that arise when the prices for hotel services are bundled. In particular, the following aspects are discussed in detail:

- Identification of the attributes within a bundle that are relevant for the consumer.
- Establishment of the bundle of services which offers the maximum utility from the point of view of the consumer.
- Determination of the degree of substitutability of the individual attributes and attribute levels that make up the package while preserving the same utility values for the buyer.
- Segmentation of consumers on the basis of the acquired sociodemographic data and elaboration of service packages that are appropriate for the identified target groups (a priori segmentation).
- Identification of target groups with similar wants and expectations in relation to the hotel service package (a posteriori segmentation).
- Measurement of the price acceptance linked to both the individual services and the bundle price.

Finding a solution to these problems is extremely important for a hotelier, since it will provide the decision-maker with valuable information as to whether his various services should be sold individually and at unit prices (unit pricing) or in the form of packages or bundles at a bundle price (price bundling).

2 On the Need for Utility-Oriented Design of Hotel Service Packages

According to fundamental marketing principles, the aim of the services offered by a supplier must be to satisfy the existing and latent wants of consumers. The requirements of the client should determine every single detail of the marketing activities initiated by a service enterprise, since it is the response of the market that in the final analysis determines a manufacturer's success or failure. At first sight, this notion of service bundle design appears easy to transform into practice. Respondents can be requested, for example, to rate the price of an overnight hotel stay on a scale consisting of “too expensive”, “just right” and “too cheap” categories. If this method of investigation was reliable, it would be possible to determine from the replies whether the price should be increased, reduced or left unchanged. This survey technique is however considered to be problematic, as interest is focused directly on the price and the real purchase situation - in which a client weighs up the price of the overnight accommodation and the utility that can be derived from it against one another - is not adequately reflected (Kucher and Simon, 1987; Wuebker and Mahajan in this book).

From the point of view of the consumer, the price represents the equivalent value of the services provided by the hotel. The consumer's conceptions of utility with regard to hotel accommodation should therefore be placed at the center of all pricing and product-policy decisions. This approach has two advantages as compared with an analysis of the respondents' individual judgments of specific facets of the service package: first of all, it is possible to determine the value attributed by the clients to each of the components that make up the bundle of services (such as the room category, catering, etc.). Secondly, the utility of the services offered by the hotel can be enhanced by means of selective modifications, thus raising the probability that more consumers will wish to take advantage of them. It is consequently crucial to establish the client's utility expectations in relation to the accommodation, in other words to measure the ratings accorded to it by the hotel guests (Simon, 1988; Wuebker, 1998).

3 Approach Adopted for an Empirical Study of Utility-Oriented Design of Hotel Services

It is clear from our deliberations so far that direct, isolated questions concerning the individual attributes of hotel services are not sufficient to facilitate utility-oriented design of service bundles and prices. On the contrary, a technique which reveals the benefits offered to consumers by a particular service package appears essential. One approach that is suitable for achieving this end is the conjoint measurement (Claxton, 1994; Barsky, 1992; Wilensky and Buttle, 1988; Wind et al., 1989; Green and Srinivasan, 1978; Nieschlag, Dichtl and Hoerschgen, 1994). This approach consists of a series of psychometric methods, the purpose of which is to determine the partial contributions of individual attributes (e.g. room category, check-in procedure) towards the formation of the global judgment (e.g. the preference for a particular service package) from empirically obtained global judgments concerning multi-attributive alternatives (e.g. the various services offered by a hotel). The presented alternatives are the result of a systematic combination of the attribute levels of several different attributes that have been identified as significant within the framework of an experimental design. Thus, we do not combine the individual, attribute-specific judgments to form a global judgment (compositional approach), but instead follow precisely the opposite procedure in extracting the partial contributions of the individual attributes and their levels from the global judgments (decompositional approach) (Bauer, Herrmann and Graf, 1995).

Table 1. Attributes and levels measured by the conjoint analysis

Attribute	Levels	
Price (P)	1. DM 175 2. DM 235 3. DM 295 4. DM 355	I II III IV
Business center (BC)	1. With secretarial services 2. With online services 3. With office equipment	I II III
Check-in (CI)	1. Check-in with personal contact 2. Quick check-in 3. Automatic check-in	I II III
Catering services (CS)	1. Exclusive, "a la carte" 2. Varied menu, "buffet style" 3. Fast, light, reasonably priced, "bistro style"	I II III
Friendliness of staff (ST)	1. Passive, "always there to help the guest (if necessary)" 2. Active, "consciously inquiring about guests´ needs"	I II
Room category (RC)	1. Budget 2. Economy 3. Standard 4. Luxury	I II III IV
Check-out (CO)	1. Breakfast check-out 2. Room check-out 3. Express check-out	I II III

On the basis of these preliminary observations, it is possible to demonstrate the suitability of the conjoint measurement for designing utility-oriented hotel service packages. This was the aim of an empirical study conducted in Munich, Frankfurt and Düsseldorf in the spring of 1996. The preferences of 86 respondents regarding the design of a service bundle offered by a hotel chain were recorded. A pilot study based on a survey of experts and clients revealed that 7 of the hotel services with a total of 22 different levels are particularly relevant for consumers (see Table 1).

The levels of the *price* attribute refer to a complete package for one overnight stay in combination with one level of each of the other attributes. The purpose of the *business center* attribute is to determine whether consumers wish to take advantage of the opportunity to make use of office equipment, secretarial services or online services. There are three possible variants of the *check-in* attribute, whereby a check-in procedure with personal contact is the most common form of service on arrival at a hotel. The quick check-in procedure is preceded by a telephone reservation initiated by the client, who is also quoted a reservation number. When he arrives at the hotel, the client is then simply asked to state his reservation number and is handed his room key by the receptionist immediately. The automatic check-in procedure likewise entails the client making a prior telephone reservation, and being given a reservation number that he types into a machine on arrival. His key is then dispensed to him automatically. Our survey attempted to establish which check-in modality offered the greatest utility to the client. The hotel guests were also able to choose between three alternative forms of *check-out*. In the case of the breakfast check-out variant, the guest is handed his check over breakfast and can also pay directly in the dining room. The room check-out procedure is controlled via the television set in the hotel room. The check is displayed to the guest on the screen. He then confirms the total amount, and the due payment is settled via his credit card. “Express check-out” means that the check is delivered to the guest by room service on the final night of his stay. The client's credit card serves as the means of payment. The *catering services* attribute represents an attempt to establish the style of restaurant that is preferred by hotel guests. The consumers were offered a choice of a buffet, a bistro or an “à la carte” menu. The *room category* attribute ranges from budget to luxury. The differences between each category concern the location of the room, its size and its furnishings. The *friendliness of staff* attribute measures the preference of clients for staff who actively inquire after their wishes and needs or alternatively for passive staff who normally remain in the background, but are there whenever the guest requires their assistance.

The data was collected in the hotel lobbies in the three above-mentioned cities. The method chosen for the study was the adaptive conjoint analysis (ACA), a computer-based technique. This investigation method is distinguished by the fact that the replies given by the individual respondents are subsequently taken into account in the next interview round. The researcher thus gradually acquires an idea of the attributes to which each consumer attaches the greatest importance. From the point of view of the participants, this interview mode means that the amount of judgment effort is reduced to a minimum. Overall, the respondents taking part in an ACA study undergo five interview rounds:

1. In the first phase of the interview, the respondent specifies the attribute levels that are most crucial to the purchase decision.
2. The respondent then places these attribute levels in a ranking order.
3. On the basis of the results of phases 1 and 2, the respondent is then asked to state the importance he attaches to the differences between two levels of each attribute.
4. In the next phase, the respondent must rate a series of product or service pairs that appear on the screen on a point scale. Each new reply causes the utility values calculated in the previous step to be adjusted accordingly.
5. Finally, the respondent is presented with a complete product design, or in this case a complete bundle of services, for which he is asked to specify a purchase probability.

The partial utility values that result for the individual attribute levels specified by each respondent are subsequently aggregated by forming the mean value and then normalized to make them more readily comparable. This ensures that the total utility value of the stimulus for which the strongest preference is expressed is 1 for all respondents. The highest, normalized partial utility value of an attribute corresponds to its relative importance (Backhaus, Erichson Plinke and Weiber, 1994). Personal data, such as the age, sex, education, profession, marital status, etc. of the respondents, was gathered prior to conducting the conjoint interviews, in order to facilitate a segment-specific evaluation of the results.

4 Results of the Empirical Study

4.1 Aggregated Evaluation of the Measured Client Preferences

The data was then analyzed in five steps:

(1) Aggregated evaluation of client preferences,
(2) A priori segmentation,
(3) Benefit or a posteriori segmentation,
(4) Market simulations, including market share calculations for competing service bundles,
(5) Calculation of a price-demand function for a specific service bundle.

Whereas the main emphasis of the first three phases was on the formation of segments (Schweikl, 1985), the focus thereafter was on a market simulation consisting of various bundles of hotel services together with a price-demand function calculation for a specific service bundle.

The individual partial utility values of each attribute level were then calculated by means of an ACA on the basis of the respondents' replies. However, suppliers are not generally interested in the preferences of individuals, but rather in the average utility values for a group of buyers (Backhaus, Erichson, Plinke and Weiber, 1994). We have refrained from presenting the individual results for this reason. The personal correlation coefficients of each respondent were nevertheless taken into account in the calculation of the average partial utility values for all respondents.

In statistics, a correlation coefficient describes the linear relationship between two attributes in the form of a number r between +1 and -1. r = +1 is referred to as a perfect positive correlation, while r = -1 is a perfect negative correlation. If r = 0, there is not a linear relationship between the two attributes (Bortz, 1993). In the context of a conjoint analysis, the correlation coefficient is a measure of the quality with which this method maps the replies of the individual respondents as interval-scaled partial utility values. All the correlations measured in our study are located in the value range from 0.63 to 1.0. The satisfactory nature of these values enabled the data of all the respondents to be evaluated. The mean, normalized partial utility values and the importance of each attribute for the total sample are shown in Table 2.

As can be seen from Table 2, the greatest relative importance (44%) is attached to the *price* when the market as a whole is considered. The highest partial utility value (0.4348) is attributed to the lowest service-bundle price (DM 175), while the highest price (DM 355) has the lowest utility value (0). The second most important attribute for the purchase decision proved to be the *room category*. This attribute contributes 21% towards the decision process regarding the choice of service bundle. The curve of the utility function has a positive gradient, in other words the better the room category the greater the utility which is derived by the hotel guest. A sharp bend appears in this curve, however, at the transition from the standard category to the luxury category. The superior, luxury room category has only a marginally higher utility value than the standard category. There is a marked increase, on the other hand, in the benefit derived by the consumer at the transition from the budget category to the economy category as well as at that from the economy category to the standard category. Thus, although consumers express a preference for the luxury room category, they are not prepared to pay a significantly higher price for it than for the economy category.

Table 2. Aggregated partial utility values for the sample

Attribute	Levels	Partial utility value	Importance
Price	DM 175	0.4348	44%
	DM 195	0.2990	
	DM 235	0.1580	
	DM 355	0.0000	
Business center	With secretarial services	0.0000	8%
	With online services	0.0780	
	With office equipment	0.0299	
Check-in	CI with personal contact	0.1712	17%
	Quick CI	0.0974	
	Automatic CI	0.0000	
Catering services	A la carte	0.0301	6%
	Buffet style	0.0625	
	Bistro style	0.0000	
Friendliness of staff	Passive staff	0.0000	3%
	Active staff	0.0317	
Room category	Budget	0.0000	21%
	Economy	0.1121	
	Standard	0.2078	
	Luxury	0.2117	
Check-out	Breakfast CO	0.0100	1%
	Room CO	0.0058	
	Express CO	0.0000	

The contribution of the *check-in* procedure towards the purchase decision is reflected by its relative importance of 17%. The consumers exhibit a preference for the personal check-in mode (partial utility value 0.1712). Guests see personal contact at the reception desk as the best means they have of influencing the service-provision process. It gives them an opportunity to express their wishes and needs, to request information about the various services that are available and to choose those that correspond most closely to their own predilections. The quick check-in procedure is also considered by hotel guests to be a valuable service (partial utility value 0.0974), while the automatic check-in option is rejected by the majority of respondents. Personal contact with the staff of the service-rendering enterprise is consequently extremely important to the hotel guests.

Considerably less importance for the choice of hotel is attached by consumers to the *business center* attribute (relative factor weighting 8%). It is conspicuous that clients appreciate the business center with secretarial services least of all, despite the fact that this is the alternative which entails the highest costs for the enterprise. Consumers derive the greatest utility from the variant with online services (partial utility value 0.0780). This reflects the already well-developed acceptance by the respondents of this relatively new mode of data communication. The provision of an office is rewarded by guests with a partial utility value of 0.0299.

A comparatively minor role in the purchase decision, namely 6%, is attributed by the interviewed persons to the *catering services*. In contrast with holiday hotels, it is in any case relatively uncommon for city hotels to bundle offers for accommodation together with meals. In spite of this, the utility values for the individual attribute levels reveal that an "à la carte" menu - the most cost-intensive variant - meets with little approval on the part of the respondents. The strongest preference is expressed for a buffet style, on account of the large number of conferences and business meals that take place in such hotels. The "bistro style" option achieves the second-highest partial utility value in the average evaluation for all respondents.

The manner in which the *staff* interact with the guests during the service-provision process is relatively unimportant for the purchase decision in the view of the consumers. It cannot be concluded from this that staff training could theoretically be dispensed with, however. With both alternatives, the respondents assume that the service from the staff will be polite and friendly at all times. The interviewed persons expressed a preference for staff who actively inquire about the needs of their guests as opposed to passive service.

The least importance is attached by the respondents to the *check-out* procedure (relative factor weighting 1%). Guests derive practically no utility from this service. It is the final element of the service-provision process that takes place within the framework of an overnight hotel stay, and is associated by the guest with payment of the check - what he sees as the unpleasant part. A tendency on the part of the respondents to favor personal service can be noted here, in the same way as with the check-in attribute. Consumers are skeptical in their attitude towards the express check-out, where they have no contact with the hotel staff. The breakfast check-out variant is preferred to the room check-out option. Table 3 shows the optimum service bundle for the average respondent, calculated on the basis of these results.

Table 3. Optimum service bundle for the fictitious average respondent

Attributel	**Levels**	**Partial utility values**
Price	DM 175	0.4348
Room category	Luxury	0.2117
Check-in	With personal contact	0.1712
Bussiness center	With online services	0.0780
Catering services	Buffet style	0.0625
Friendliness of staff	Active	0.0317
Check-out	Breakfast CO	0.0100
Total utility value		**1.0**

4.2 The Implications for Price Bundling

If we assume that a consumer is willing to spend more money on a service from which he derives a greater utility, his acceptance of the prices for various services can be inferred from the partial utility values of the various attribute levels and the relative factor weighting of these attributes. The services relating to the business center and catering are of particular interest for the purposes of our study, since they are not presently offered to clients of the hotels at the three locations in question within the framework of price bundles.

These two services play a relatively minor role in the respondents' purchase decision in favor of a particular bundle of services, with a relative importance of 6% for the catering and 8% for the business center. Price acceptance is likely to increase in line with relative importance. A consumer will be more willing to pay a higher price for a better category of room, for example, than for a business center fitted out with all the latest equipment. Thanks to the linear price-utility function, it is possible to calculate the change in price acceptance for a step up from a less strongly preferred attribute level to a more strongly preferred one (Bauer, Herrmann and Graf, 1995). The procedure is described below with the aid of an example.

If we assume that the hotel enterprise would like to offer a buffet style instead of a bistro style, the increase in clients' willingness to pay (in other words in their price acceptance) can be calculated as follows. The difference between the utility of the two attribute levels is 0.0625. The utility value that ensues if the price of the room is raised from DM 175 to DM 355 is 0.4348 lower. On the basis of this calculation, and owing to the linearity of the price-utility function, the willingness of consumers (x) to pay for individual attributes is the result of a simple proportionalization of the partial utility values:

$$\frac{180}{0,4348} = \frac{x}{0,0625} \quad => \quad \frac{180}{0,4348} \cdot 0,0625 = x => \quad \mathrm{x} = \mathrm{DM}\ 25.87.$$

The consumer is willing to pay DM 25.87 more for the buffet style. Table 4 lists the monetary differences for various supplementary services.

Table 4. Price acceptance of various service enhancements by the fictitious average respondent

Attribute	Service bundle enhancement	Utility difference	Monetary difference
Business center	From "secretarial services" to "office equipment"	0.0299	DM 12,37
(importance 8%)	From "office equipment" to "online services"	0.0481	DM 19,91
	From "secretarial services" to "online services"	0.0780	DM 32,29
Catering services (importance 6%)	From "bistro style" to "à la carte"	0.0301	DM 12,41
	From "à la carte" to "buffet style"	0.0324	DM 13,41
	From "bistro style" to "buffet style"	0.0625	DM 25,87
Room category (importance 21%)	From "budget" to "economy"	0.1121	DM 46,40
	From "economy" to "standard"	0.0957	DM 39,32
	From "standard" to "luxury"	0.0039	DM 1,61
	From "budget" to "standard"	0.2078	DM 85,41
	From "budget" to "luxury"	0.2117	DM 87,64
	From "economy" to "luxury"	0.0996	DM 40,98

The above propositions are confirmed by the calculated price acceptance values. With regard to the attribute with the higher relative factor weighting, the consumer will generally also display a greater average price acceptance in the event of a change from a less strongly preferred attribute level to a more strongly preferred one. The average price acceptance if the room category is raised from one step to the next is DM 50.21, for example. This contrasts with an average price acceptance of DM 21.52 if the business center is enhanced by replacing a less strongly preferred attribute level with a more strongly preferred one. The same correlation is shown again in Fig. 1, which also takes account of findings concerning the distribution of price acceptances within a maximum-price space (Adams and Yellen, 1976; Simon, 1992; Wuebker, 1998).

It can be seen that the price acceptance values for the respondents are located in a marginal region of the coordinate system. It can be concluded from this that simple price bundling will lead to a lower profit than unit pricing (Simon, 1992). Mixed price bundling can also be an advantage however. Consumers who set greater store by the business center and the catering services, or who have only failed to use them in the past due to a lack of awareness, can be motivated to make use of these services through mixed price bundling and the price saving they achieve as a result.

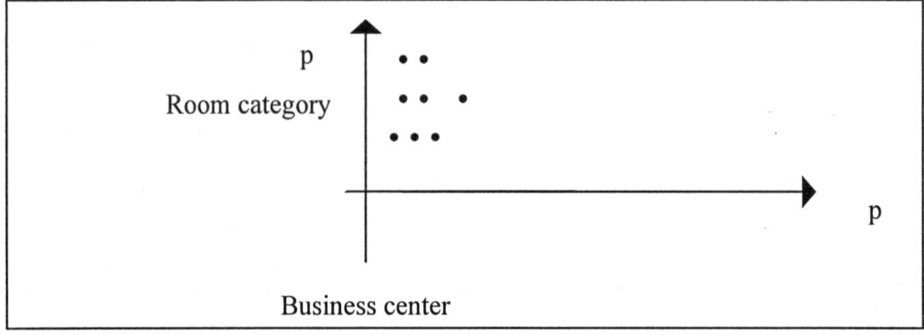

Fig. 1. Price acceptance values for the fictitious average respondent

4.3 A Priori Segmentation

In addition to identifying and attending to customers' wants, full implementation of marketing principles also entails subdividing the heterogeneous overall market into several homogeneous segments. The reasons for this step include the need to communicate with a specific target group or to apply a segment-specific pricing policy. Aggregated partial utility values were therefore calculated for target groups that had previously been defined a priori on the basis of the personal data gathered from the respondents. In view of the restricted scope of our work, only a few, selected results are outlined in brief below for the predefined market segments.

(1) A priori segmentation according to sex

Price is the most important aspect of the purchase decision for both men and women, whereby the price-utility function has a negative gradient. Women are only marginally less sensitive to price than men. Sex-specific differences were revealed in connection with the check-in procedure and the room category. The check-in mode is twice as important for men than for women (18% relative importance as compared with 9%). On the other hand, female guests attach greater importance to the room category (30% as compared with 20%). Moreover, it is possible to discern an increase in the partial utility values for the female segment with every step up in the room category. In contrast, a reduction in utility was established for the group of male respondents at the transition from the standard category to the luxury category.

(2) A priori segmentation according to age group

An analysis of the data based on the age of the respondents revealed that younger guests in particular (under 30) are relatively insensitive to price, and that they attach considerable importance to the business center attribute, with a relative factor weighting of 20%, yet consider the check-in procedure to have almost no relevance for the purchase decision (relative factor weighting 5%). The differences between the other age groups were only minor.

(3) A priori segmentation according to profession

The differences between self-employed persons and the other two groups (salaried employees and managers) merit somewhat closer attention. The entrepreneurially active place greater importance on the room category and the catering services. They are moreover less sensitive to price than salaried employees and prefer an "à la carte" menu. In addition, the self-employed prefer the hotel staff not to inquire actively about their needs, but rather expect them to be able to anticipate their wishes automatically. The preferences of the other two groups are extremely similar.

(4) A priori segmentation according to the reason for staying in the hotel

The calculated partial utility values and the resulting factor weightings indicate that private clients are much less sensitive to price than business clients. People who are not traveling on business attach greater importance to the catering services, whereby an “à la carte” menu is preferred to a bistro or buffet style. It would appear that private guests are in a position to devote more time to enjoying their meals. These guests accordingly also set greater store by the hotel staff (13% relative importance as compared with 6%). In contrast with business clients, private guests' ideals do not include an active hotel staff.

(5) A priori segmentation according to the frequency of hotel stays

The results obtained by segmenting according to the frequency of hotel stays also appear to have a certain practical relevance. The message of the acquired data is absolutely clear: the more frequently a person spends the night in a hotel, the less significant price becomes and the greater the importance attached to the room category. Moreover, the business center attribute - and in particular the alternative with the online services - is extremely relevant for travelers who stay in hotels constantly. Whereas the “often in hotel” and “rarely in hotel” groups still prefer the traditional check-in mode with personal contact, those respondents who stay in hotels regularly expressed a preference for the quick check-in variant.

(6) A priori segmentation according to brand loyalty vis-à-vis the hotel enterprise

Crucial differences were revealed in relation to the price and room category attributes between regular clients and those guests who were residing in the hotel purely by chance. Regular clients who are already familiar with the various alternative room categories are also willing to pay a higher charge for a better category. Whereas, for example, the price acceptance of regular clients at the transition from the budget to the economy category is DM 57.63, that of non-regular clients is a mere DM 36.05. As far as the other services taken into account by the study are concerned, the differences between the two segments are only minimal.

(7) A priori segmentation according to hotel location

The segment of participants who were interviewed in Frankfurt is characterized by an above-average relative factor weighting for the room category (30%). It is particularly striking that, while these respondents still derive a significant utility gain from a transition from the standard room category to the luxury category (positive utility difference of 0.0753), the other two groups suffer a utility loss. A comparison of the Munich and Düsseldorf respondents reveals that the Düsseldorf guests pay greater attention to the friendliness of the staff than those in Munich (relative factor weighting of 9% as compared with 1% in Munich).

4.4 Benefit Segmentation

In contrast with segmentation based on sociodemographic criteria or observable consumer behavior, a posteriori segmentation (Gutsche, 1995) (benefit segmentation (Yankelovich, 1964; Becker, 1993)) entails forming groups according to the individual importances (partial utility values) determined by means of the study. The 86 respondents were subdivided into five groups, according to the similarities between their individual partial utility values, with the aid of a hierarchical cluster analysis. The following typology was obtained for the hotel guests, taking account of all the above-mentioned variables:

Cluster A: The price-conscious, traditionally-minded clients

The members of group A are characterized by a well-developed price consciousness (47%). It is noticeable that the price-utility function flattens out at the transition to the second attribute level (DM 235). The group members are thus relatively insensitive to prices in the upper regions. The second most important attribute for the members of this group is the room category (17%), whereby these guests tend to prefer the standard alternative. In line with their traditional attitudes, they also prefer a check-in procedure with personal contact, passive staff and a business center with secretarial services. The catering services and the check-out procedure play only a subordinate role for this consumer segment. This cluster, which embraces 27 members, is the largest out of the five and the one that exhibits the greatest degree of similarity with the aggregated results for the complete set of respondents. It is distinguished by a large number of clients who frequently spend the night in a hotel (78%), whose professional training is based on an apprenticeship (27%) or who are employed in the retail sector (18%).

Cluster B: The miserly clients

The members of this group are extremely unwilling to pay a higher price for a quality enhancement within the service bundle. The cluster differs from the other four groups above all in that it accords the highest relative factor weighting to the price (54%) and attaches the least importance to the room category (6%). The high weighting of the price attribute has the effect of reducing price acceptance in relation to the other services. The steep downward gradient of the utility function as the price increases suggests a decline in the consumers' willingness to purchase. The availability of a business center is also very important for this cluster, whereby - in contrast with all the other groups - the greatest store is set by the provision of office equipment. This reflects the low quality aspirations of these respondents. The catering services play a similarly significant role for these hotel guests (12%). On the other hand, services such as the check-in and check-out procedures and the friendliness of the staff are relatively unimportant to such consumers. As expected, this cluster contains a relatively large number (67%) of non-regular clients, since loyal clients are usually satisfied consumers and as such distinguished by a higher price acceptance (Meffert, 1994). The persons in this segment only seldom spend the night in a hotel (40%), which explains their low quality aspirations.

Cluster C: The staff-oriented consumers

This group differs from the other segments above all in the great importance it attaches to the friendliness of the hotel staff towards their guests. The relative factor weighting of the "friendliness of staff" attribute is 26%, causing this attribute to rank above price (21%) as a purchase-decision criterion. It is extremely important to these clients that the staff inquire about their needs and make an active contribution towards the success of the service-provision process. This is reflected by the curve of the utility function for the "check-in" attribute, which is the third most important service for this group: the more frequent the contact with the staff during the check-in procedure, the stronger the preference for these attribute alternatives. With a relative importance of 11%, the business center plays a major part in the choice decision. The huge importance of the staff factor can be explained primarily by the high proportion of non-regular clients (71%) in this cluster. These guests are not yet familiar with the peculiarities of the hotel and are thus dependent on the assistance of its staff. In addition, the majority of those belonging to this cluster only rarely spend the night in a hotel. This factor also contributes towards the special role allotted to the staff.

Cluster D: The snobs

This is the second largest group out of the five clusters. This group of consumers sets particular store by the room category (32%) when choosing a service bundle. The fact that the snobs attach less importance to price than any of the other clusters (19%) is a further remarkable characteristic. On the other hand, they accord a greater relative importance than all the remaining four clusters to the check-in and catering services attributes (20% and 16% respectively). It is noticeable that these guests generally prefer the more cost-intensive services, especially a check-in procedure with personal contact and an exclusive, à la carte menu. Owing to the relatively low partial utility values for the price, the price acceptance of these consumers with regard to the - in their opinion - superior alternatives of the other attributes is comparatively high. They are willing to accept a price of DM 297.45 in return for a step up from the budget category to the luxury category, for example. As far as the passive segmentation variables are concerned, the high proportion (45%) of guests who stay in hotels constantly is conspicuous. These predominantly young consumers (50% under 40 years of age) are apparently not prepared to forego luxury when they stay in a hotel. This group is moreover distinguished by having the highest proportion (60%) of regular clients. It is thus clear that such clients - and especially those at the Munich hotel - have high quality aspirations.

Cluster E: The innovators

Cluster E is the smallest group in the study. The important factors for these hotel guests are the price (24%), the business center (23%), the room category (17%) and the check-in procedure (15%). With regard to the room category, it is noticeable that these clients are satisfied with standard equipment (highest utility value = 0.1654). In addition, it is possible to ascertain a positive attitude towards new, innovative services. The business center with online services is given preference by the innovators over the other two service types. As far as the check-in procedure is concerned, these consumers again display their openness vis-à-vis new ideas by preferring the “quick check-in” and “automatic check-in” variants. An analysis of the background variables reveals that these guests stay in hotels constantly, or at least frequently. It is very important to such consumers to be able to communicate with their environment from the hotel, as reflected by the high utility value of the business center with online services. They also see the hotel as simply a place to spend the night. The service-provision process is only of secondary importance to them. They want to be finished with the unavoidable services, such as the check-in and check-out procedures, in the shortest possible time. This segment is moreover characterized by a high proportion (57%) of regular clients. These clients are already familiar with the hotels and therefore have no need of personal contact when they check in.

4.5 Market Simulations with Different Service Bundles

The results of the conjoint analysis have already been presented and interpreted in the above sections. Following on from this, the individual partial utility values can be used to conduct simulations of the respondents' choice patterns in relation to various service bundles, and these simulations in turn taken as a basis for forecasting market shares (Stadtler, 1993). The first step in estimating these shares is to define a series of alternative service bundles that will compete with one another on a market. In addition, the simulation is founded on the assumption that these few alternatives are the only ones available on the market. Table 5 shows the competing service bundles defined for the purposes of our study.

Table 5. Competing service bundles

Attribute	**Bundle A** Level	**Bundle B** Level	**Bundle C** Level
Price	DM 175	DM 235	DM 355
Business center	Office equipment	Online services	Secretarial services
Check-in	Quick CI	Automatic CI	Personal contact
Catering	Bistro style	Buffet style	A la carte
Staff	Passive	Active	Active
Room category	Economy	Standard	Luxury
Check-out	Express CO	Room CO	Breakfast CO

Service bundle A represents a cheap alternative, in which the quality of the attribute levels is correspondingly lower. The supplier of this bundle is aiming to establish himself in the segment of price-oriented buyers. Bundle C is the antithesis of bundle A. It is expensive, but the attribute levels it contains are of a high quality. Bundle B corresponds to a market participant who is attempting to become established as a supplier of services somewhere between these two extreme positions.

The maximum utility choice rule (MUC) (Mengen, 1993), also known as the first choice rule (Balderjahn, 1994), is used in this study to calculate the market shares. The underlying idea behind this method is that, in arriving at a purchase decision, a consumer chooses the bundle of services from which he derives the greatest overall utility. Consequently, the total utility values for each of the three service bundles on offer were calculated for all the respondents on an individual level. The total utility value for respondent no. 2 is 0.8628 for bundle A, for example, 0.7903 for bundle B and 0.1381 for bundle C. According to the MUC rule, this consumer would choose bundle A in a simulated purchase decision. All in all, 39 out of the 86 respondents prefer bundle A when measured in this way, while 21 would purchase bundle B and 26 bundle C. The resulting market shares are shown in Fig. 2.

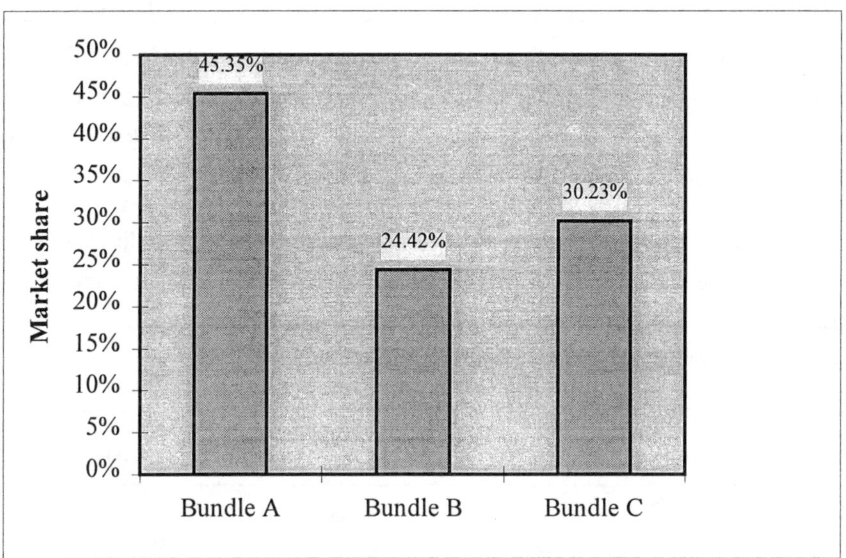

Fig. 2. Simulated market shares of service bundles A, B and C

Despite the relatively poor quality of its attribute levels, the largest market share is achieved by bundle A. In second place is the expensive - but high-quality - bundle C. The lowest market share is obtained by bundle B, with 24.42%. The next question is how to vary bundle B in order to increase its market share. The following simulations provide an answer.

Variations of bundle B with no reactions by competitors

Bundle B can be varied in several different ways. The variations with a positive effect on the market share of bundle B are shown in Table 6 (Stadtler, 1993). All other variations result in a lower or identical market share. Bundles A and C remain constant for all the simulations.

Table 6. Variations of bundle B with a positive effect on the market share

Variation within bundle B From (initial situation)	to (new attribute level)	Market share	Difference
DM 235	DM 175	40.70%	+ 16.28%
BC with secretarial services	BC with online services	31.40%	+ 6.98%
Automatic check-in	Check-in with personal contact	43.02%	+ 18.60%
Automatic check-in	Quick check-in	31.40%	+ 6.98%

The only way to achieve a significant increase in the market share would be either to introduce a check-in procedure with personal contact to replace the automatic check-in or to reduce the price. The shifts in the market share that would result from these measures are shown in Fig. 3.

It can be seen that both these isolated changes to bundle B would result in an increase in the market share to a level well above that of the competing bundles. A particularly marked improvement in the market position of bundle B could be achieved by introducing a new check-in procedure.

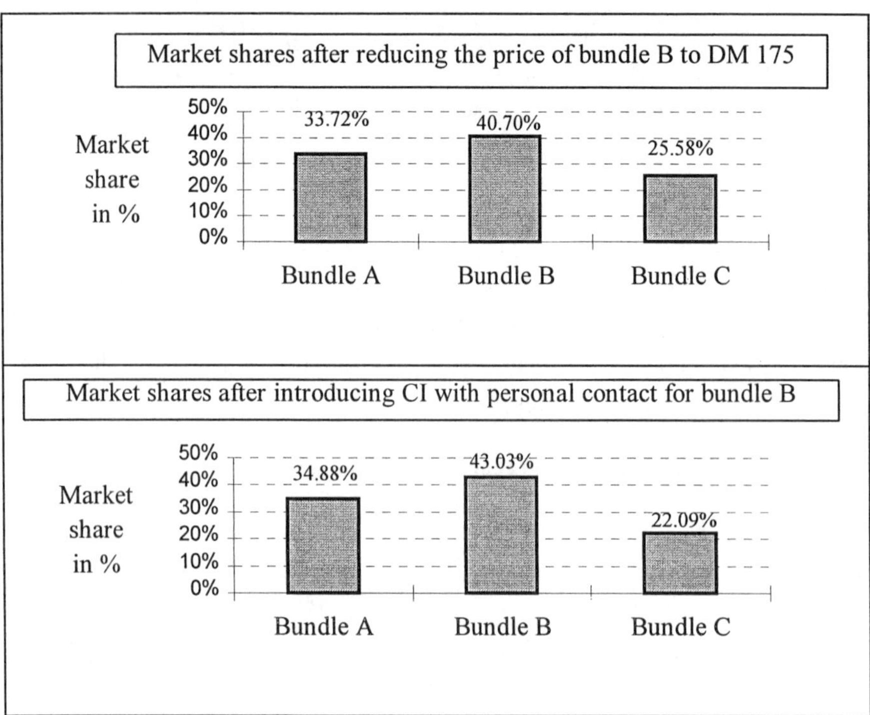

Fig. 3. Shifts in market shares with no reactions by competitors

Variations of bundle B with reactions by competitors

The calculation of the above market shares is based on the assumption that the services offered by the other two market participants remain unchanged. It is more likely, however, that these enterprises will react to the substantial drop in their own market shares by varying their ranges of services as well. Table 7 shows one variation of bundle A and one of bundle C in response to the introduction of a check-in procedure with personal contact in bundle B.

Table 7. Shifts in market shares with reactions by competitors

Market situation	Bundle A	Bundle B	Bundle C
Initial situation	45.35 %	24.42%	30.23%
Variation		Introduction of CI with personal contact	
New situation	34.88 %	43.02 %	22.09%
Reactions	Introduction of CI with personal contact		Price reduction from DM 355 to DM 295
Final situation	40.70%	29.07%	30.23%

By likewise introducing a check-in procedure with personal contact, the supplier of bundle A is able to win back his original market lead. Nevertheless, this hotelier is obliged to accept a drop in his market share of 4.64% as compared with the initial situation. The supplier of bundle B can restore his original market share if he reduces his price by 60 monetary units. Bundle B has managed to improve its position on the market by 4.64%. The overall winners of this scenario are the consumers, who obtain either service bundles offering a greater utility for the same price (A and B) or an equivalent service bundle for a lower price (bundle C).

4.6 Calculation of a Price-Demand Function for a Specific Bundle of Hotel Services

The high relative factor weighting of the price, both within the individual segments and on the market as a whole, reflects the importance of the role played by the pricing strategy. If, as in our study, price is one of the attributes evaluated within the framework of a conjoint analysis, it is possible to calculate a price-demand function with the aid of the individual partial utility values (Balderjahn, 1994; Simon and Kucher, 1988). An attempt is made below to calculate this function for bundle C, whereby it is assumed that the three service bundles described above are the only competing ones on the market, since this bundle corresponds most closely to the ideal service bundle of the fictitious average respondent (apart from the price and the level of the catering services attribute). The calculation of the price-demand function is based on the maximum utility choice rule. Assuming that the other two service bundles remain constant, the price of bundle C is varied for all the different attribute levels and compared with bundles A and B. Table 8 shows how many respondents prefer bundle C to bundle A and bundle B for various prices, according to the first choice rule.

Table 8. Demand curve for bundle B

Price of bundle C	Quantity demanded	Market share
DM 175	59	68.60 %
DM 235	50	58.14 %
DM 295	38	44.19 %
DM 355	26	30.23 %

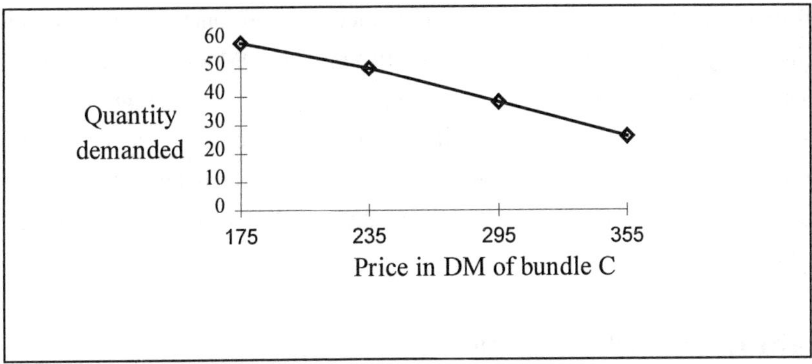

Fig. 4. Price-demand function for bundle C

Fig. 4 shows a negative, almost linear price-demand function (Diller, 1991; Simon, 1992). A small bend is visible in the function where the price changes to DM 235. Reducing the price from DM 235 to DM 175 has less effect (increase of 9 bundles in the quantity demanded) than reducing it from DM 355 to DM 295 or from DM 295 to DM 235 (increase of 12 bundles in the quantity demanded). The price elasticity (Simon, 1992)

$$\varepsilon = \frac{percentage\,demand\,change}{percentage\,price\,change}$$

is exactly 0.703 (+18% / -25.6%) for a price reduction from DM 235 to DM 175, whereas the value for a price reduction from DM 355 to DM 295 is -2.7 (+46% / -17%).

To sum up, it can be stated that the effect of a change in the price of bundle C on the quantity demanded (sales) is greatest in the upper price bracket. The final step of our analysis was to calculate a price-demand function for each of the three hotels in which the survey was conducted, the aim being to identify the potential for differential pricing according to region (Diller, 1991).

Since all three price-demand functions have very similar curves, it was not possible to derive any concrete information that might be relevant for regional price differentiations. It was however revealed that - for an identical price - bundle C achieves a larger market share in Düsseldorf than in Frankfurt, and a larger share in Frankfurt than in Munich. The price-demand function for the *Munich* guests is approximately linear, i.e. the market share declines by the same relative amount each time the price is increased. The price-demand function for the hotel guests in *Frankfurt* also has a linear curve up to a price of DM 295. An increase in the price above this figure is however accompanied by a rise in the guests' price sensitivity, as the negative gradient of the function becomes steeper at the transition to this attribute level. The persons interviewed at the hotel in *Düsseldorf* are characterized above all by an increase in their price sensitivity between the two middle price levels.

5 Summary and Outlook

The conjoint analysis method was used to determine which individual services play the greatest part in the decision to purchase a particular bundle of hotel services. In addition, assertions were made regarding the prospects for success of various service packages. It also proved possible to determine the price acceptance of consumers in the event of a change from a service with a lower utility to one with a higher utility. Implications for price bundling modalities were derived from these findings.

Subsequent to an a priori segmentation of the respondents, and the resulting identification of differences between them, the hotel clients were subdivided into five different target groups with the aid of a cluster analysis. In the final part of this paper, the alternative service bundles were compared in terms of their market opportunities and a price-demand function was calculated for one of them. There are many marketing questions to which this type of approach is unable to provide adequate answers, however. If the hotel manager needs to know which of the product-policy measures derived from the conjoint measurement are most suitable for maximizing the profits of his enterprise, for example, he must also examine the consequences of these activities for his cost structure. An approach that allows the costs of the different attribute levels to be integrated in the analysis is evidently essential (Bauer, Herrmann and Mengen, 1994). The "demand side" of the enterprise must therefore be linked to the "cost side", in order to determine the product capable of yielding the maximum profit.

References

Adams, W.J. and J.L. Yellen (1976). "Commodity Bundling and the Burden of Monopoly." Quarterly Journal of Economics, Vol.90 (August), 475-498.

Backhaus, K., B. Erichson, W. Plinke and R. Weiber (1994). Multivariate Analysemethoden. 7th edition, Berlin.

Balderjahn, I.. "Der Einsatz der Conjoint-Analyse zur empirischen Bestimmung von Preisresponsefunktionen." Marketing ZFP, Vol. 1, 12 - 20.

Barsky, J. D. (1992). "Customer Satisfaction in the Hotel Industry: Meaning and Measurement." Hospitality Research Journal, Vol. 16, 1, 51-73.

Bauer, H. H., A. Herrmann and A. Mengen (1994). "Eine Methode zur gewinnmaximalen Produktgestaltung auf der Basis des Conjoint-Measurment." Zeitschrift für Betriebswirtschaft, 64 Jahrgang, 81- 94.

Bauer, H. H., A. Herrmann and G. Graf (1995). "Die nutzenorientierte Gestaltung der Distribution für ein Produkt." In: Jahrbuch der Absatz- und Verbrauchsforschung, 41 Jahrgang , 4-15.

Becker, J. (1993). Marketing-Konzeption: Grundlagen des strategischen Marketing-Management. 5th edition, München.

Bortz, J. (1993). Statistik für Sozialwissenschaftler. 4th edition, Berlin, Heidelberg, New York.

Claxton, J. D. (1994). "Conjoint Analysis in Travel Research: A Manager`s Guide." Travel Tourism, Hospitality Research: Handbook for Managers and Researchers, Ritchie, J.R.B./Goeldner, P. (ed.), 2nd edition, 513-522.

Diller, H. (1993). "Preisbaukästen als preispolitische Option." Wirtschaftswissenschaftliches Studium, 22. Jahrgang, Vol 6, 270-275.

Diller, H. (1991). Preispolitik. 2nd edition, Stuttgart.

Dolan, R. J. (1987). "Managing the Pricing of Service-Lines and Service-Bundles." In : Wright, L.K. and M.A. Cambridge (Ed.). Competing in a Deregulated or Volatile Market, Report No. 87-114, Marketing Science Institute.

Goldberg, S., P. E. Green and Y. Wind (1984). "Conjoint Analysis of Price Premiums for Hotel Amenities." Journal of Business, Vol. 57 (January), 111-132.

Graumann, J. (1983). Die Dienstleistungsmarke. München.

Green, E. P. and S. Tull. (1982). Methoden und Techniken der Marketingforschung. 4th edition, Stuttgart.

Green, P. E., A. M. Krieger and M. K. Agarwal (1991). "Adaptive Conjoint Analysis: Some Caveats and Suggestions." Journal of Marketing Research, Vol. 28 (May), 215-222.

Green, P. E. and S. Srinivasan (1990). "Conjoint Analysis in Marketing: New Developments with Implications for Research and Practice." Journal of Marketing, Vol.54 (October), 3-19.

Guiltinan, J. P. (1987). "The Price Bundling of Services : A Normative Framework." Journal of Marketing, Vol. 51 (April), 74-85.

Gutsche, J. (1995). Produktpräferenzanalyse : Ein modelltheoretisches und methodisches Konzept zur Marktsimulation mittels Präferenzerfassungsmodellen. Berlin.

Kucher , E. and H. Simon (1987). "Conjoint Measurement : Durchbruch bei der Preisentscheidung." Harvard Manager 3, 28-36.

Meffert, H. (1994). Marketing-Mangement: Analyse, Strategie, Implementierung. Wiesbaden.

Mengen, A. (1993). Konzeptgestaltung von Dienstleistungsprodukten: Eine Conjoint-Analyse im Luftfrachtmarkt unter Berücksichtigung der Qualitätsunsicherheit beim Dienstleistungskauf. Stuttgart.

Nieschlag, R., E. Dichtl and H. Hörschgen (1994). Marketing. 17th edition, Berlin.

Wilensky, L and F. Buttle (1988). "A Multivariate Analysis of Hotel Benefit Bundles and Choice Trade-Offs." International Journal of Hospitality Management, Vol. 7, 1, 29-41.

Wind, Y., P.E. Green, D. Shifflet and M. Scarbrough (1989). "Courtyard by Marriott: Designing a Hotel Facility with Consumer-Based Marketing Models." Interface, Vol., 19, 1, 25-47.

Authors

Adam, Richard is Director of Marketing at the Arabella Hotel Group, Munich, Germany.

Bauer, Hans H., Ph.D., is Professor of Marketing and Business Management and Head of the Institute of Marketing at the University of Mannheim, Mannheim, Germany.

Cornet, Andreas, graduated engineer, is a Consultant in the Duesseldorf office of McKinsey & Company, Inc., Germany. He is a member of the Assembly-Industry-Sector and a Consultant in the field of steel- and automotive industry and product development.

Coulter Higie, Robin, Ph.D., is an Associate Professor of Marketing at the University of Connecticut, Storrs, Connecticut.

Fine, Charles H., Ph.D., is Professor of Production Management at the Sloan School of Management at the Massachusetts Institute of Technology (MIT). Furthermore he is Co-Director of the International Vehicle-Program. His scientific interests are quality management and flexibility in production.

Fuerderer, Ralph, Ph.D., Adam Opel AG, International Technical Development Center, Ruesselsheim (Germany). He received his Ph.D. degree from the Koblenz School of Corporate Management, Vallendar (Germany) and is author of Option and Component Bundling under Demand Risk.

Herrmann, Andreas, Ph.D., is Professor of Marketing and Business Management at the University of Mainz (Germany). He is member of the advisory board at Ctcon, Management Consulting, Vallendar (Germany).

Huber, Frank, doctorial student in Marketing at the chair of Professor Hans H. Bauer, University of Mannheim (Germany). He studied business administration and economics in Mannheim (Germany) and London (England).

Huchzermeier, Arnd, Ph.D., is a Professor of Production Management at the WHU in Koblenz, Otto-Beisheim-Graduate School, Vallendar, Germany. He held lectures at the Universitys of Chicago, Pennsylvania, Western Ontario and Vienna as well as at INSEAD.

Lingnau, Volker, Ph.D., is an Assistant Professor of Production Management and Control at the Business School at the University of Mannheim, Germany. He holds a master's degree in industrial engineering from the Berlin Technical University and a doctor's degree in business administration from the same university. His dissertation was about the management of product variant. He also works as a business consultant for international companies especially in the field of management accounting and management control systems.

Mahajan, Vijay, Ph.D., is the John P. Harbin Centennial Chair in Business, Graduate School of Business, The University of Texas, Austin, Texas, 78712. He received a B.S. in Chemical Engineering from the Indian Institute of Technology, Kanpur, India, and a M.S. in Chemical Engineering and Ph.D. in Management from the University of Texas at Austin.

Monroe, Kent B., D.B.A., is the J.M. Jones Professor of Marketing and Head of the Department of Business Administration, University of Illinois, Champaign-Urbana, Illinois. He has pioneered research on the information value of price and authored the leading text Pricing: Making Profitable Decisions. Monroe is a member of the Advisory Board of the Pricing Institute, and a Fellow of the Decision Sciences Institute.

Neumann, Carl-Stefan, Ph.D., is a Partner in the Frankfurt office of McKinsey & Company, Inc., Germany. He earned his Ph.D. degree in Physics and has worked at McKinsey since 1990 in the fields of netbased industries, strategy and information management.

Ringbeck, Juergen, Ph.D., is a Partner in the Duesseldorf office of McKinsey & Company, Inc., Germany. He got degrees in business administration and in mathematics and works at McKinsey since 1988 in the field of strategy, growth and information management. Ringbeck together with Duesseldorf director Rolf-Dieter Kemois is leading the international research project "Excellent Information Management in the Production Industry".

Schrage, Linus, Ph.D., is a Professor of Production Management at the Graduate School of Business, University of Chicago. He has had previous or visiting positions at Stanford, UCLA, Pontificia Universidade Catolica/Rio de Janeiro, Universidad Torquato di Tella/Buenos Aires, and Universite Catholique Louvain-la-Neuve/Belgium. He has a book on simulation with Ben Fox and Paul Bratley. He has several books on applied optimization. His research and publications have been on the subjects of production planning and logistics. He also played a founding role in the development of the widely used optimization packages: LINDO, LINGO, and the What's Best spreadsheet optimizer.

Simon, Hermann, Ph.D., is Chairman and Chief Executive Officer of SIMON, KUCHER & PARTNERS in Bonn, Germany, and Cambridge, Massachusetts. He is also a Visiting Professor at the London Business School. Simon is the author of Price Management, Hidden Champions and Power Pricing.

Suri, Rajneesh, Ph.D., is an Assistant Professor of Marketing at the College of Business Administration in Drexel University, Philadelphia. He earned his Ph.D. in Marketing at the University of Illinois at Urbana-Champaign, Illinois.

Tönshoff, Nils, Ph.D., is Project Coordinator for concepts of production and controlling at the BMW AG, Munich, Germany.

Wricke, Martin, doctorial student in Marketing at the chair of Professor Andreas Herrmann, University of Mainz, Mainz, Germany. He studied business administration in Mainz.

Wuebker, Georg, Ph.D., is a Consultant in the European Office of SIMON, KUCHER & PARTNERS in Bonn, Germany. He studied business administration and economics in Hull, Great Britain, Osnabrueck, Germany, and Austin, Texas. He received his Ph.D. degree at the University of Mainz and wrote his dissertation on bundling.

Yadav, Manjit S., Ph.D., is an Associate Professor of Marketing at the Department of Marketing, Mays College & Graduate School of Business, Texas A&M University, Texas. He obtained his Ph.D. in marketing from Virginia Polytechnic Institute and State University. His research focuses on firms' pricing strategies, consumers' price perceptions, and electronic commerce.

Druck: Strauss Offsetdruck, Mörlenbach
Verarbeitung: Schäffer, Grünstadt